SELECTING COLOUR
FOR PACKAGING

CUMBRIA COLLEGE OF ART & DESIGN LIBRARY

Three Week Loan

*This book must be returned on or before
the date last stamped below.*

Fines will be charged for items returned late.

Selecting Colour for Packaging

E P Danger

Gower
Technical Press

Published by
Gower Technical Press Ltd.,
Gower House,
Croft Road,
Aldershot,
Hants GU11 3HR,
England

Gower Publishing Company,
Old Post House,
Brookfield,
Vermont 05036,
U.S.A.

British Library Cataloguing in Publication Data
Danger, E.P.
        Selecting colour for packaging.
        1. Colour in packaging
        I. Title
        688.8 TS195.2

ISBN 0-291-39716-6

Printed by Dotesios (Printers) Ltd,
Bradford-on-Avon, Wiltshire

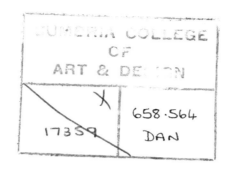
Contents

Contents

Preface

Although colour is not the only factor in achieving good packaging, there will be little argument with the axiom that the effective use of colour is an important marketing tool and is the key to packaging that sells. However, the effective use of colour as part of the graphic design of a package presupposes selection of the right colour or colours, and this book is intended to help those who have to make that selection and thus ensure that a package has maximum impact in the market place.

The plan has been to bring together the best available information on the subject and to provide a source of reference to the characteristics of colour in the marketing and packaging concepts. The result will, it is hoped, answer three basic questions:

- Should colour be used at all?
- Which colours should be used?
- How should colour be used to best advantage?

These questions are faced by all those responsible for the supply and design of packaging including users of packaging, designers, advertising agents, marketing experts and management generally. The questions are also faced, although less directly, by manufacturers, producers and converters of packaging and packaging materials, not forgetting those who sell packages in a range of standard colours. It is believed that all these categories will find help and interest in the pages that follow.

Selection of the most effective colour for packaging does not mean following a few simple rules, or using guesswork, or relying on the prejudices of individuals. It is a complex process which embraces marketing considerations, design considerations and practical considerations before the physiological, psychological and optical aspects of colour can be applied to a specific situation.

It is not advisable to make any final decisions about colour until the nature and shape of the package has been determined. Colour, by itself, will not make a poor package into a good one; packaging design is a total concept and any package has to be designed in a positive way. To that end, the designer needs to be furnished with an adequate brief setting out all the factors that have to be taken into account.

This is not a manual of packaging, and it does not attempt to discuss the advantages or disadvantages of different materials or different types of pack. The general principles of package design are discussed in a broad way, together with those considerations that should be taken into account when preparing a brief for the designer, because the same considerations are equally applicable to the selection of colour.

The principal emphasis has been placed on packaging for new consumer products because this is where colour is most important and consequently where the marketing aspects of packaging have priority. However, if a package is to be successful and to play a positive part in the marketing process, the functional and other aspects of the package must not be forgotten and must be included in the design brief. These aspects are just as important as the visual aspects and they have been listed and discussed where appropriate. Sections dealing with changes in existing packaging, and with the packaging of industrial products, have also been included.

This book has been planned to bring together the best available information, and the material has been divided into five parts as follows:

- Part I deals with packaging in general terms, its function and the principles that add up to the ideal package. It also includes a section dealing with packaging assessment and notes on changing packaging. The material included may be described as background to packaging.

- Part II sets out the principles governing selection of colour for packaging applications, explains the nature of trends and summarises the factors that should govern colour selection. Perusal of this part will explain the reasons for the recommendations made in the remaining parts.

- Part III lists the various points that should be included when preparing a brief for the designer and includes notes on the effect of each point on colour selection. It forms a comprehensive checklist of the many factors that have to be taken into account when designing a package and which have to be reviewed when preparing a colour specification.

- Part IV underlines those points which require special attention when preparing a design brief or colour specification for the packaging of specific types of products such as cosmetics or food. It also includes a note on the packaging of industrial products.

- Part V describes the mechanics of colour selection and outlines the preparation of a colour specification that will pinpoint the various attributes and characteristics of colour needed to achieve a successful result. This part also includes a comprehensive colour index from which it is possible to select individual hues having the right characteristics.

- Finally, there is a selection of sources of further information, and an index to all the products mentioned in the book.

Parts I and II may be described as background material which helps the reader to understand the principles involved. Parts III and IV are the planning stage, where marketing and other factors are brought together and used to identify those colour characteristics which are essential to a satisfactory outcome. Part V outlines the colour selection process.

During the course of some thirty years experience as a colour research consultant, I have frequently been asked to suggest sources of information about the use of colour in packaging. Hardly a week goes by without letters and calls from budding marketing executives who have been assigned the task of writing a thesis on the subject of colour and packaging and are frantically seeking basic information on the subject.

My answer to these enquiries is that I do not know of any source to which easy reference may be made. Although the subject is frequently discussed in the trade and technical press, the articles usually have a limited objective and are not easy to track down. So far as I am aware there is no one source of information which is at all helpful. This book is an attempt to fill that gap.

The material in the book is based on my own files, researches and practical experience and reflects knowledge acquired in making colour recommendations to many well known companies. It also includes the results of an extensive study of published information dealing with both colour and packaging and the interface between the two; this study has extended over many years.

In addition, I have had the inestimable advantage of access to the files and experience of Faber Birren, the world's leading authority on colour, whom I have been privileged to call a friend and associate for many years. I owe a deep debt of gratitude to Mr Birren for permission to use his material when writing this book.

E.P. Danger
London
April 1986

Part I
GENERAL PRINCIPLES

1 About packaging

1.1 The functions of packaging

One dictionary defines packaging as the design and manufacture of packages for retail commodities, but it is a good deal more than that and applies equally to consumer and industrial products. Packaging has come to mean the art or science of using a package as a marketing tool, and that is the meaning intended in this book. Packaging is a complex subject which has become an essential part of the promotion of any product, although particularly consumer products, and it cannot be separated from selling.

Even in the Stone Age there were undoubtedly packages because things had to be carried around and protected from frost and rain, and a package remained a functional concept for many centuries. It is only in this century that the package has become something more and that the concept of packaging has become an essential part of the economics of existence. There are two main reasons for this. The first is that a more complex civilisation and higher standards of living make it necessary to have more elaborate packages in a functional sense. The second is that the package has become an essential part of the process of selling. This is due partly to the efforts of the packaging industry in creating new ideas and partly to the development of package design; together, these have created something which did not exist before. However, there have also been changes in the nature of retail trade, such as the development of supermarkets, which depend on packaging.

In order to understand the changes it is desirable to draw a distinction between packaging and packages, although the two are often treated as being synonymous. Packaging embraces the whole concept including the immediate package, the outer, the wrapper and so on, and the part that the whole plays in marketing and merchandising. A good package will not sell any product unless the packaging concept is right, nor will it sell a poor product. A poor package may give a poor image to an otherwise excellent product, however well thought out the packaging concept may be.

If packaging is to be used to maximum advantage in the marketing process, the individual package must perform a number of vital functions. It must

- protect the product and keep it in good condition
- provide a convenient and economical means of distribution
- be cost effective
- sell the product.

These functions have developed over the years. Originally a package was simply a handy receptacle from which a shopkeeper dispensed handfuls of tea or which prevented peas from rolling on the floor. As time went on, the distribution system became more complex and consumers insisted on higher standards of hygiene; it became essential for products to reach the purchaser in good condition and the need for adequate protection during the distribution process became increasingly important. Individual wrapping became vital when cartons and outers were used to facilitate handling and began to play a part in the display of the product.

The second function became important as channels of distribution lengthened and costs of distribution became a heavier burden. In addition, of course, methods of trading changed radically and packaging had to adapt to self-service conditions and to the complications of distribution through multiple outlets.

Costs involved in packaging include warehousing, efficiency of operations, capital costs of packaging machinery, costs of inners and outers, costs of packing on to pallets, freight and many other factors. Any packaging has to justify itself economically either by improving rates of production and reducing wastage; or by providing easier and cheaper distribution and therefore lower handling and storage costs; or by making the product more saleable, thereby reducing unit costs. Ideally, the package should contribute to all these.

All these functions are just as important as they ever were, but today the selling function is even more important. This is because of the pattern of modern trade, especially self-service methods of selling. An increasing number of products are expected to sell themselves and maximum sales cannot be achieved unless the product can be seen by the ultimate purchaser – and seen under the best possible conditions. In many cases sales depend on the image created by the packaging.

Packaging provides the link between the promotional support given by the manufacturer, the shelf space allocated by the retailer, and the needs and desires of the ultimate purchaser. For this reason, packaging has to be considered at every stage of the marketing process and the packaging concept has become a vital part of marketing. However, all the thought in the world devoted to packaging will be a waste of time if the individual package does not appeal to the ultimate consumer. Appearance is particularly vital.

1.2   The packaging manufacturer

There are four parties mainly involved in the concept of packaging:

The packaging manufacturer, who designs and makes packages but who is primarily concerned with *selling packages*.

The user of packaging, namely the company that puts a product into a package and that is primarily concerned with *selling the product*. The user is responsible for marketing.

The distributor or retailer, who influences the nature of the package and who is the object of that part of marketing which is known as merchandising.

The customer, the ultimate purchaser of the product contained in the package, who is the object of the whole concept and to whom the package must appeal.

Although this text is intended primarily for the user of packaging, it must discuss the company that designs and makes packages because it play an important part in the packaging concept; it develops and creates new packaging forms which must inevitably affect the thinking of those responsible for packaging policy. There must be, or should be, close co-operation between packaging users and packaging manufacturers, and the continued success of the latter depends in the long run on the success of the packaging user.

The user of packaging must be alert to changes in packaging technology, to the development of new forms of packaging and to changes in the price of materials, and it is part of the function of the packaging manufacturer to bring them to the attention of interested parties. New developments are often initiated by packaging manufacturers who see a gap in the market and in some cases they react to changes in the nature of trade and to the creation of new products that require packaging. In other cases the packaging manufacturer may be briefed by users to produce a new package, for example plastics cans for motor oil.

The packaging manufacturer can change the whole marketing picture by developing new packages or methods of packaging. The modern self-service store would never have developed without present day concepts of packaging which have, in turn, developed from efforts to cut the cost of retail distribution. The disappearance of the small trader made the advent of self-service operations inevitable, and co-operation between users and manufacturers of packaging provided the means of achieving the objective.

The development of new forms of packaging is of course a continuous process, and some packaging manufacturers, such as Metal Box, have not only experimented with new types of packaging like plastics cans but have also developed existing containers by experimenting with types of food which suit them. Another development by Metal Box has been standard plastics bottles in various sizes with various options for distinction. For example, the bottles can be moulded in any colour and texture can be added; alternative caps can also be in colour; the bottles can be printed or can carry labels, and they can be fitted with inserts to sprinkle or squirt the contents. A standard bottle of this nature reduces tooling costs and makes for economy in packaging. They are also immediately available and are of particular benefit to small companies which would otherwise be faced with heavy initial costs.

All these are examples of packaging manufacturers taking the initiative in a marketing sense.

1.3 Marketing

It is the user of packaging who is the important party in the context of this book and who gains by using packaging as an integral part of marketing; the functions of the individual package (listed in Section 1.1 of this part) all contribute to the ultimate objective of a satisfied customer. Packaging is equally important to the marketing of industrial products, although there may be different emphasis on the various functions.

Ensuring that a complex piece of machinery reaches its destination in good condition is part of good marketing because it ensures a satisfied customer, but the package does not sell the product in the same way that a novel pack sells an Easter egg.

Packaging is the art or skill of using a package to maximum advantage at all stages of the marketing process. To appreciate the connection between packaging and marketing it is first necessary to be certain that the meaning of marketing is understood. Marketing is often used as a synonym for selling but it is a good deal more than that. There are many definitions of marketing, but almost all of them agree that it is a strategy which starts with the conception of a product and ends with the sale of the product to the ultimate user and therefore embraces production, packaging, distribution, selling, merchandising and point of sale activities.

This broad definition covers the whole complex pattern of design, production, distribution and selling by which we live. At each stage of the strategy packaging may have a part to play; it certainly has to be considered, and it may dictate the course of action to be taken.

The extent to which packaging is concerned with all the various stages depends on many factors including the nature of the product, the availability of raw materials, the results of research, the theme of advertising and the scope of the marketing plan. Packaging is a great deal more than just a pretty container and the ultimate justification for taking trouble over it is customer satisfaction which ensures repeat sales. This is also the ultimate objective of the whole marketing process.

The relationship between packaging and marketing divides into a number of parts:

Marketing planning The overall strategy which includes the production of goods, the way that they are sold and the place of the package in the selling process.

Merchandising The operation which includes getting the product to the point of sale safely and which ends with the sale of the product.

Customer attitudes These often have an influence on the course of action to be taken.

The appeal of the package to the customer This ultimately makes the sale.

Marketing planning must take into account the link between the package, the sales theme, promotion, advertising, and the various other components of marketing. The basic ideas behind the packaging of a £1000 appliance are not very different from those behind a £1 lipstick.

Some commentators believe that packaging will overtake other forms of advertising as a marketing tool before the turn of the century; economic conditions have increased the fundamental role of packaging as a sales aid and packaging users are increasingly experimenting with packaging as a means of highlighting their brands against alternatives. Some users fail to treat packaging as part of promotion; it should not be determined by technical considerations or left to a junior brand manager.

Some reports, however, suggest that packaged goods marketing has run its course. Proctor & Gamble are reported to believe that the area of consumer wants and desires has been so thoroughly combed over that there are no obvious needs waiting to be filled. Each firm is playing roughly the same game so that no one ever gets a clear cut advantage. This is, of course, an argument in favour of more thought and care in the design of packaging. The public is still willing to pay for packaging but marketing demands that the package should be distinctive.

The importance of packaging as a primary promotional tool means that marketing people need to know about production lines as well as about packaging as a communications medium. It is necessary to strike a balance between technical and functional considerations and what the brand manager wants the packaging to do, including appeal to the consumer visually and in handling. In addition, there must be a knowledge of reproduction processes.

Marketing management also needs to understand cost factors, especially where a new package is involved; it would be suicide, for example, to consider soft drinks in cans when the manufacturer has just installed a new bottling plant. The argument usually centres on unit cost versus exclusivity, but technology is not always the answer and old ideas may be best.

1.4 Merchandising

The third party involved in the concept of packaging is the distributor or retailer, and appeal to the retailer is an important part of the marketing process. Before any package can perform its selling functions it has to secure display space on the shop counter or shelves; this means that the product, *and its package*, must appeal to the retailer in a number of different ways. The retailer has to be persuaded that shelf space allocated to the product will justify itself in sales.

This part of the marketing process is generally termed merchandising. This is the operation that ends with the sale of the product, but which involves getting it to point of sale safely and with the goodwill of the retailer and ensuring that it is seen and identified by the customer. Merchandising may also be described as the struggle for display space and the whole packaging concept has to be sold to the retailer first of all. Unlike the customer, the retailer may be more interested in ease of handling and convenience of storage and will certainly not be interested in damaged goods or faulty packages. However, the retailer also needs to be persuaded that the package will sell the product.

The struggle for display space also includes the management of discounts and good representation, as well as good packaging, and in the present day concept of merchandising the principal task of the manufacturer's representative is to secure shelf space and to see that the product is adequately displayed. Special offers, dump displays and similar techniques play a part in the struggle to get the product within the view of the customer, and in all this packaging has a vital part to play.

Merchandising techniques have extended from the grocery trade to confectionery, cosmetics, pharmaceuticals and many other sectors; virtually every manufacturer uses the same techniques and provides the same displays, racks, etc. and for this reason the package plays a key role. A novel package may persuade the retailer to allocate shelf space but novelty tends to wear off and something more basic is required. The retailer may also be willing to feature products that have heavy promotional support as long as the promotion is continued, but it is the whole packaging concept which secures continued goodwill and ensures consistent display.

All packaging must lend itself to display at point of sale and even packaging cases and outers should add something to the appeal of the store. The manufacturer can no longer depend on the retailer taking an active part in the selling of a product unless it is fast moving. This applies particularly to urban areas where site values are high; a fast turnover in as small a space as possible is essential for survival.

Retailers will favour those products which require the minimum of handling and selling and packaging helps to achieve this objective. Whatever the type of store, the retailer needs to display as much as possible to maintain turnover and will be unfriendly to a product which sticks on the shelves or which needs positive effort to sell.

1.5 Customer attitudes

The customer is the fourth party involved in the conception of packaging, and the ultimate objective of the whole marketing process is to sell a product to the customer. The relationship between packaging and marketing is affected by broad customer attitudes, and at the consumer level there is often a negative attitude to the whole idea of packaging; this attitude has spurred the development of own brands, generics and other changes at retail level. These attitudes should be clearly understood by marketing management because they may influence overall planning.

Most ordinary people associate packaging with consumer products and there is often a misunderstanding about what it is or does. In some quarters packaging is associated with switch selling and other undesirable practices. Indeed, there is a body of opinion which considers that it is undesirable to do anything to create demand and hence, because good packaging does or should create demand, that packaging is undesirable. This body of opinion also believes that packaging adds unnecessarily to the cost of the product and that consequently the product should be sold with the minimum of wrapping.

These viewpoints are illogical. We live in a complex industrial environment whose prosperity depends, in the long run, on selling more goods to an ever increasing population which expects an ever increasing standard of living. This depends on creating demand, thus ensuring incentives to greater production. Also, incidentally, depressed demand for packaging materials causes industry to suffer. There is very good reason for doing everything possible to increase demand for consumer products and for the packaging in which they are sold. It benefits everybody.

Moreover, the idea that packaging adds to the cost of a product is also illogical in most cases. Packaging often makes for economics in handling and distribution and saves storage space, thus making the product cheaper. The packaging may also help to create demand which does not already exist, thus leading to increased production and ultimately to lower unit costs. There may be a case for using one form of packaging rather than another because it will save natural materials, but this is no argument against using any packaging at all. A number of world organisations have been built up on the demand created by good packaging and these organisations contribute to the prosperity of the countries in which they operate.

Americans spend substantial sums per year on packaging, and each of them is said to bring home about 300 kg (600 lbs) of packaging in a year. In fact, it has been estimated that as much as nine cents in every dollar spent went on packaging but this money was not necessarily wasted; the container protected the goods and provided product information. Although cans or boxes are more expensive than bags or pouches, they save labour and shelf space. Packaging is partly a form of promotion but it also fulfils the desire of the consumer to handle their foods and beverages in convenient sizes. The way in which a society packages its food and other products reflects the quality of its life and, although the way a product is packed appears to be a secondary consideration compared with price, in fact the container influences the consumer's choice and is a silent salesman.

In recent years social attitudes to packaging have become important and economists have suggested that social pressures may force an overall reduction in the use of packaging per product unit. Various interest groups, such as those concerned with the environment, have criticised certain aspects of packaging and there is a general feeling that packaging must be socially acceptable. The industry has been anxious to demonstrate its responsibility by recycling materials wherever possible. Environmental pressures may have the effect of increasing the cost of packaging.

Comment in the women's press tends to attack the use of packaging to create shelf impact and to call for more 'useful' packages and for greater use of recycling, but at least one feature has come to the conclusion that, although convenience has to be paid for, it may be worth it. Much of the criticism by consumer organisations shows a lack of knowledge of the reasons for packaging; most consumers would not want to go back to the pre-packaging era. Major manufacturers have nothing to gain by misleading customers; competition is fierce and requires constant review of both the packaging concept and its costs, so that any package that was misleading or too costly would soon be forced out of the market.

Packaging is a product of the revolution in retailing that has replaced the assistant behind the counter; it can be overdone but in general the cost is part of the economics of mass production and excessive cost will soon force change. The consumer is protected by a vast fund of laws and regulations covering labelling, description of contents, weight of contents, and similar matters. In fact, there is a risk of the consumer being overprotected; the multiplicity of the regulations may add unnecessarily to the cost of the product. The British Packaging Council has published a code for consumer packaging which has the following points:

- Packaging must comply with legal requirements.
- The package should give adequate protection to contents.
- The materials used should have no adverse effect on the contents.
- The package should not be a misleading size, or leave voids.
- The package should be convenient for handling by the customer.
- The package should convey all relevant information.
- There should be due regard to the environment.

The task of those responsible for packaging is to strike a balance between offence to consumer bodies who think there is too much packaging; offence to customers who want more convenient packages and more product protection; and offence to retailers who are primarily interested in packages that are easy to handle. Striking the balance requires careful attention to design.

1.6 Package appeal

The fourth element in the relationship between packaging and marketing is the appeal of the individual package to the customer – the factor that ultimately makes the sale. This is primarily a matter of the design of the package, which has to communicate the marketing story that the manufacturer wants to put over, fulfil the needs of the retailer, reflect the current attitudes of the consumer, and have visual appeal made up of form, shape and colour which catch the eye of the customer and trigger that action which leads to a sale.

People seldom buy products for the sake of the package itself, except perhaps confectionery (such as chocolates) and cosmetics (such as scent). They buy the product because of the message that the package conveys to them, and it follows that package design requires a great deal of thought and is by no means a simple matter.

In the case of a new product, the package has to stimulate and educate the customer; it may have to excite curiosity and inform the customer about the benefits of the product; it may have to persuade the customer to break existing habits and try something new; and it may have to perform many other functions depending on the nature of the product. Above all, the package must convey a product identity. In many cases packages represent the brand in the eyes of the consumer, and when people recall the brand they visualise the package; in such cases the package makes the sale. In other cases, the package is an adjunct to sales and makes the product itself more visible or more attractive, or perhaps displays it to better advantage at point of sale.

To achieve an ideal package, the whole subject has to be approached in a logical and systematic way, starting with the marketing plan and proceeding by way of research to a brief for the designer. The process starts with a number of key questions:

- What part does the package play in the marketing plan and what does the package have to do?
- What type of package will best suit the circumstances and what is the most economical form of packaging?
- What shape, or form, will be most suitable?
- What colour, or colours, will achieve maximum impact?

2 The ideal package

2.1 The approach

The individual package is the key element in the concept of packaging and may be a box, a bottle, a bag or some other form of container which holds a product. The major part of this guide inevitably concerns the package that appears on the retail shelf, but the term *package* also includes any variations such as immediate wrappers which may be purely utilitarian in nature, and cartons or outers which may have a display function as well as a distribution function. This point should be remembered when reading this section.

The individual package, whatever its nature, helps packaging to fulfil its proper role in the marketing process, and therefore it must perform the same functions as packaging as a whole and outlined in the previous section. The ideal package must be effective in protecting its contents, it must make a realistic contribution to effective distribution, and it must be cost effective in the sense that its costs are not out of proportion to its benefits. However, the most important function is the part that it plays in selling. The ideal package must be successful from a functional point of view, from an economic point of view and from a selling point of view at one and the same time. For example, it should be easy to handle, easy to open and good looking.

When planning a new package it is a good idea to take a cold, clear look at the whole packaging situation from a number of different viewpoints, notably

- marketing
- production and distribution
- cost accounting
- management in general.

All the viewpoints must be combined together to achieve the primary aim, which is to sell the product contained within the package. The package is a vital part of the marketing mix and may actually create the sale; a box of chocolates, for example, is often bought primarily for its package and the contents may be a secondary consideration. With products such as convenience foods, the package is an essential part of the proposition. The package must appeal to the natural outlets for the product and these should be defined in the marketing plan. There cannot be a successful package without a marketing plan which clearly sets out what the package has to do.

Unfortunately, the importance of the marketing plan is often overlooked and too much emphasis may be placed on the production and distribution functions; the package may protect the contents so well that the purchaser cannot get at them. Or, there may be too much emphasis on economy, and although the package is cheap and easy to produce it lacks sales appeal. Sometimes the package reflects the prejudices of an individual, such as the chairman, who knows very little about packaging and who may place too much emphasis on graphic design so that other attributes are sacrificed for the sake of a pretty picture.

A good package has to be a compromise between a number of conflicting viewpoints; there may be a need to resolve differences of opinion and to recognise that a package must have both practical appeal and sales appeal. This important point is not always appreciated, and a few years ago a US trade mission drew attention to a fundamental difference in attitudes towards packaging in the US and the UK. In the US the average manufacturer considers the package to be a vital marketing and merchandising tool, but in the UK it is often considered to be primarily a means of protection and distribution.

2.2 Planning

Planning a successful package involves three essential elements:

The marketing plan This defines the place of the package in the overall marketing plan and governs everything else because it sets out, or should set out, the characteristics of the market at which the product is aimed and to which the package must appeal. Product and packaging must be co-ordinated from the time that a new product idea develops, and it is a great mistake to try to design a package as an afterthought. The nature of the package may have a vital effect on product pricing; a higher performance package may justify a higher price for the product. The package may also have to be designed to fit in with existing production machinery or with packaging machinery that is readily available.

Functional design This includes choice of suitable material, the way that the package protects its contents and the way that it fits into the production and distribution flow. This may be called the technical design of the package, but it should also include practical appeal to the consumer in the form of ease of opening, convenience in use, etc.

Graphic design These are the visual factors that attract the attention of the consumer and help the package to make a sale. The term includes shape, form and colour and the graphic design on the package. A well designed package should have visual appeal, act as a miniature poster and be an eye catcher and shopper stopper; the package should also have sensory appeal, although this may also be part of functional design.

Each of these elements is just as important as the other; a good package is more than just a pretty picture. Both functional and graphic design must be governed by the marketing mix to ensure that the package is suitable for the market that it has to reach.

It follows from the above that the first step in the creation of the ideal package is to consider the marketing plan and, although each package is different and should be considered individually, there are a number of vital features that experience and a study of the market place have shown to be important in most cases:

- The package must be appropriate to the product contained in it, in marketing, functional and graphic terms. This underlines the importance of co-ordinating package and product.
- The package must be appropriate to the class of trade for which it is intended. The marketing plan must identify the market and graphic design must appeal to it.
- The package must be up to date in both functional and graphic terms and should reflect the latest marketing position. Packages that have been on the market too long lose sales and the marketing plan should predicate periodical review.
- The package must be functionally well designed for its target, fixed in the marketing plan. This means that it must protect the product, be easy to open, be easy to handle, be easy to stack, and be economical in the situation in which it will normally be sold.
- The package must be attractive to, and acceptable to, the retailer in marketing, functional and graphical terms.
- The package should lend itself to display and be of a shape and size which makes it easy to see on the shelf. This requires a combination of marketing policy, functional design and graphic design.

- The package must be well designed graphically. Graphic design is a creative function aimed at a specific target, fixed in the marketing plan, and should ensure that the package commands the eye, invites picking up, and appeals to the ordinary man or woman. Colour plays an essential part.
- The package should identify the product and its brand image where appropriate; it may also have to establish a link with other products of the same manufacturer where marketing policy so dictates. Graphic design, typography and colour are an integral part of the marketing plan.
- The package must appeal to the majority, and not the minority, in most cases. The graphic design should reflect current trends except where marketing policy dictates a different approach.
- The package must communicate and tell consumers what they are buying. Functional design and graphic design must interpret the marketing message and put it over.
- The package must stand out from, and be different to, the competition. This presupposes a study of the competition and its packages. Again, graphic design and functional design must interpret a marketing policy.

2.3 Marketing aspects

The first step in achieving the ideal package is consideration of the marketing plan and the place of the package in that plan. The conception and design of the package must reflect the marketing policy of the packaging user. The marketing plan may be simple enough in the case of a basic product which only requires a wooden packing case, but it may be extremely complex and sophisticated in the case of a competitive consumer product.

In either case, however, it is necessary to follow through the whole process of marketing from the conception of the product to the ultimate user, and each stage must be reflected in the design of the package. At each stage, management must decide its course of action and may need to use market research to discover the needs of the market place or to test the reactions of consumers and retailers to the ideas that they propose to use.

Obviously the package has to be appropriate to the product, and this must be the first consideration. However, if the package is to play its proper role it must have customer appeal, and this means that all concerned must have a clear idea about the type of customer that they are trying to attract; there has to be a clear definition of the market and what the package is supposed to do in that market. Every package has to be designed in a positive way. You cannot stick a product in any old bag and hope for the best, but nor can you design a package in a vacuum. The package must be aimed at a specific market and the preferences and requirements of that market studied in detail before there can be any attempt at design; this study should encompass the factors that are likely to influence the sale of the product in that market.

All too often a package design is produced without any clear idea of the potential market. A good product in a cheap, badly designed package may never appeal to the market that is its normal outlet; on the other hand, a package that is too elaborate may not stand up to the handling that it receives in a busy supermarket although it may be just right for the speciality store. The nature of the prospective customer is also important. Children, or the parents of children, will not be attracted by a package that nearly cuts off their fingers when they try to open it. The likes and dislikes of children can make or break a product and so may the concern of parents for the welfare of their children.

Some manufacturers do not have any real idea about the make-up of their markets and the motivation of their customers. Some have very little idea how their packages are used and what function they perform before sale and after sale. Without this knowledge it is hardly possible to create a successful package. Design must reflect the job that the package has to do. Many packages are left until the last moment and then cobbled together as an afterthought; worse still, the package may be a compromise between the conflicting views of a committee, none of whose members has any clear idea of aim. Sometimes neither the designer, nor any one else, has sufficient knowledge of marketing to appreciate the needs of the market place.

Another important point which is essentially part of marketing is how up to date the package should be. There are cases where a traditional approach is appropriate, but in most cases the package should be up to date in terms of both functional design and graphic design. This point is perhaps of more importance when considering the future of an existing package, but no design should give an impression of being dated.

Once the broad marketing picture has been pinpointed, the production and distribution process can be studied to discover the conditions that are likely to be met before the product reaches the consumer. Compromise may be necessary. An ideal solution from a market planning viewpoint may not be practical with existing packaging machinery; there may be shortages of, or difficulties with, packaging materials; the handling that the package receives in transit may dictate a particularly robust construction; a small package may be ideal from a marketing angle but impossibly expensive to produce. Many other similar factors need to be considered.

2.4   Functional aspects

There is a functional aspect to the whole packaging chain which includes the container itself, the inner wrapper or liner, immediate wrappings, outer carton, packaging case and so forth and the materials of
which each is made. A package may be perfectly satisfactory for its purpose and have an excellent wrapper, but it would have failed in its purpose if it was contained in an unsuitable outer which did not stand up to
handling or which did nothing for the image of the product; if the outer
is visible at point of sale it may just as well have some display value.

   If packaging is to fulfil its role in the marketing plan, it must be functionally sound at all stages. It follows that design cannot start until there
is a final marketing plan which specifies whether the product is to be
sold in bulk, in plastics bags or in a box. In some cases the nature of the
packaging will be dictated by the product process; frozen foods are a
typical example, but with confectionery a number of alternatives are
possible with comparatively small changes to the production line.

   Thus, the real starting point for the functional design of an individual
package is the marketing plan, but in practical terms it begins with the
production line, is affected by the whole distribution process and takes
into account the needs of the ultimate customer. Functional design
starts with the premise that the package should be suitable for its purpose. and it should be constructed of materials which are appropriate to
the product and the nature of the handling expected. It follows that the
process should reflect knowledge, and experience, of materials; knowledge of methods of manufacture; familiarity with channels of distribution, experience of the way that retailers handle and store the package;
and knowledge about how the package is used by the ultimate purchaser.

   Materials may, of course, be changed from time to time, perhaps because existing materials become too expensive or are in short supply or
perhaps because of change for the sake of change. New types of packaging are continually being developed and there may be a case for trying a
new type, or trying novelty, to liven up demand. A new type of package
may have innate consumer appeal, whatever its appearance; the 'carry
home' pack for soft drinks and the like was a case in point that filled a
real need and provided an example of packaging manufacturers taking
the initiative to solve the marketing problems of users.

   On the other hand, many novel packs have severe faults in functional
design and never gain acceptance once the novelty has worn off. Ease of
handling at retail level needs particularly careful consideration, especially with supermarket sales. Packages which cause the retailer a great
deal of trouble to handle, store and price will lose goodwill. The finicky
and fiddling display outer which is difficult to set up and gets in the way

of everything else will hardly be given pride of place. Under present day conditions the retailer has considerable influence over the functional design of packaging and the larger multiples are often in a position to dictate the form of packaging and the way that it is distributed. Some insist on pallet loads of individual units or in packing in outers of a type which facilitates handling in the store. These requirements may, in turn, affect the shape and size of the individual package.

Functional design must also take into account the likes and dislikes of the consumer. Few customers will be attracted by a package that will not go into their shopping baskets because it is an awkward shape; nor will they repeat buy a package that breaks up and spills the contents as soon as they get it home. Ease of opening is particularly important and any difficulty can cause considerable consumer resistance; someone who has tried to open a can of sardines without a key may never buy sardines again. Most people get annoyed if they cannot get the wrapper off a toffee or if they break the biscuits when they take them out of the pack, or if they cannot empty a package completely. Failure to consider these points can destroy the most carefully thought out marketing plan.

The appearance of the product within the package is also important and a functionally sound package will enhance appearance.

2.5 Graphical aspects

The principal function of graphic design in the production of the ideal package is to attract the attention of the customer (normally the consumer) and, having attracted that attention, to create a sale. Many purchases are made on impulse because customers do not know (broadly speaking) what they want when they reach the point of sale; one of the aims of graphic design is to arouse this impulse. In some cases it is the graphic design which makes the sale; most boxes of chocolates, for example, are sold because of the design on the box rather than for the contents of the box, although a well established brand will usually sell better than a less well known one.

The appeal of a package may, of course, have nothing to do with visual impressions or graphic design. Appeal to the consumer comprises a number of factors, some practical, some functional, some physiological, some psychological. For example, people have become conditioned to handy packs and easy to carry packages and these can be a potent contribution to sales irrespective of design. However, although all packages must appeal to the consumer in both functional and marketing terms, all the convenience in the world will not persuade people to buy something that looks unpleasing.

A pleasing appearance is important to all customers but it is particularly important where the consumer – the ordinary retail purchaser – is concerned, and it is on the consumer that all parties to the packaging concept depend for their living. The ultimate aim of packaging is to make a sale, and the actions of the consumer at point of sale will be guided by a number of considerations:

- what they see
- the visual attraction of what is seen
- the image created by what is seen, recognised because it has been promoted.

The ideal package is one that meets all these considerations and it is the graphic design that ensures that it has maximum appeal. If consumers do not like what they see then they will not buy, and this may happen when a package is designed primarily to protect its contents and visual appeal is forgotten. All the technical skill embodied in the package will not persuade consumers to buy something that does not look good. A package that may be ideal from a production point of view has failed in its purpose if it does not have visual appeal.

3 Customer appeal

3.1 Visual attraction

Every package must have customer appeal if it is to do its job properly, and this is vital in supermarket selling where the key is self-selection and the package must sell direct to the customer off the shelf. It must appeal to the eye and make the customer want to reach out and take it; it must give the customer confidence; it must sell at speed; and it must ensure that products sell themselves. Customer appeal is a combination of a number of factors, including

- visual attraction
- shape
- form
- colour
- trends.

There is often a conflict between academic design and what the customer will buy, and the design must reflect the needs and reactions of ordinary people. People as a whole react subconsciously to certain stimuli and the great mass of people are motivated by the same considerations which must be reflected in design.

The most important factor in achieving customer appeal is visual attraction, and this is quite independent of the nature of the product or such attributes as subtlety of flavour. Visual attraction lies in the subconscious and comprises basic desires and feelings modified by changing tastes and prejudices. While it is fairly easy to discover whether a package is functionally attractive or whether a particular method of packaging commends itself to the consumer, it is much more difficult to discover whether the package is visually attractive until it is proved by sales. There is little point in asking consumers what they would like because few can visualise what would appeal to them – just as they cannot foresee what action they will take at point of sale.

The solution to the problem is to make use of basic desires and reactions. This does not mean designing a package according to a set of rules, and nothing takes the place of the creative skill of the designer. It does mean taking advantage of the fund of knowledge about the psychology of the average consumer and about the physiological and optical appeal of design. An appealing package must reflect the principles of human perception and put them to use. Evaluation of a package by the individual goes far beyond the eye; it involves the brain and the human psyche itself.

The work of psychologists has a direct and practical bearing on package design because it identifies the whys and wherefores of human reactions and behaviour. Application of the basic principles of *Gestalt* psychology is indicated. The package makes an impression on the mind going far beyond what the camera records, and package design needs to take this into account alongside functional and mechanical considerations.

If the package is to have maximum appeal it must make an instantaneous impression which is simple and direct; this is especially important in self-service conditions when the exposure of the package to the eye is of very short duration. Other points:

- Simplicity and regularity of design are recommended.
- The impression of the package should be pleasing from both distance and short view.
- The package should be easy to recognise for what it is; if the customer has to peer to understand it, then the design is not doing its job.
- People will not try to decipher a design; keep it simple.
- Packages should have a clear identity and not merge into one another on the shelf.

The basic stimuli that help to create a visually attractive package are shape, form and colour, all of which should be up to date and appropriate to the market. These stimuli are discussed in more detail in the sections that follow.

3.2 Shape

The shape of a package is one of the basic stimuli that help to create a visually attractive whole. However, there can be no fixed principles governing the physical shape of a package because it is usually dictated by the nature of the product; by mechanical considerations; by selling conditions; by display considerations; and by the way that the package is used, for example it may be carried around in the pocket.

However, there are some basic rules which should be followed whenever possible:

- Simple shapes are preferred to complicated ones.
- A regular shape will have more appeal than an irregular one and the latter may cause a mental blockage which impels the customer to turn to something else. An irregular shape for a novelty package may have virtues on occasions but generally speaking the average person prefers something simple.
- A shape which is not balanced will be unpleasing.
- Squares are preferred to rectangles and a rectangle which has a square root is preferred to one that does not. This point is of maximum importance when the packages are grouped together on a shelf.
- Shapes should be tactile and soft.
- A convex shape is preferred to a concave one; the convex shape has good tactile qualities which invite picking up and conveys a friendly atmosphere. The Coca Cola bottle is the perfect commercial illustration of this point; it combines the best of both worlds by having both a convex and a concave surface. The combination of the two is almost irresistible because people, and especially children, have a tendency to reach out for things that are soft and convex and which may be held snugly in the hand. The bottle also, incidentally, resembles the human female form.
- Round shapes are preferred by women and they like circles better than triangles. Angular shapes are preferred by men and are considered more masculine; men prefer triangles to circles. These sexual connotations may be useful in some cases, depending on the nature of the product.
- Shapes should be easy on the eye; some shapes are distorted in the angles of peripheral vision.

The shape of a package certainly has an influence on sales and experience shows that changing the shape of a package makes a difference. Shape is also part of design, but it is suggested that marketing needs should take preference over purely visual considerations when deciding shape, and research may be useful. Some points that have been discovered include the following:

- A man's hat would be more acceptable in a hexagonal box but a woman's hat is best in a round box.
- Toilet soap for women sells best in an oval shape.
- Detergents have been found to sell better if they convey mas-

culinity even though they are bought by women; a square box is indicated.

3.3 Form

In the present context the word 'form' refers to the design that appears on the package itself or on its label, and many of the basic rules that apply to shape, and which are set out in the previous section, apply equally to form. The design should achieve proportion and balance and should be simple and regular because the eye and the brain prefer it that way; anything that is askew tends to bother the eye and to create a mental block so that the impact of the package is lost. At the time of writing this book most people are attracted by smooth, flowing lines rather than by hard, angular ones because this is a trend of the times. However, there are design trends just as there are colour trends and preferences may change.

Some broad rules which apply are as follows:

- For optimum appeal, do not cut the design in half. Quite frequently the overall design is cut in two by a brand name or line, and this causes the package to look less attractive and smaller than it actually is. The same effect can be caused by contrasting colours.
- Form should be tactile and soft, like shape. No package should be so 'hard' that people are subconsciously repelled from picking it up, and this may be particularly important with packages that are bought primarily by women.
- Do not present sharp angles. Elements in a design that are too sharp should be toned down, especially if the primary appeal is to females.
- Contrasts of light and dark should not be too marked. They may make the package too difficult to 'grasp' and may break the design into small parts which create the wrong effect.
- Some forms are apt to become distorted in the angles of peripheral vision.
- Some forms have unpleasant associations and should be avoided.

3.4 Colour

Colour is the most important of the basic stimuli which create visual attraction and customer appeal and it is an essential part of the graphic design of a package. The use of colour is central to the whole process of package design but it must be used with purpose and not simply for the sake of colour. When selecting colour for packages it is desirable to consider first the basic principles of perception, and then colour for products, markets and selling conditions.

Colour is important because it appeals to the emotions and not to reason. People first notice the colour of the package; impressions of shape and form come much later, but before the product in the package is noted, or the print on it. Colour is particularly vital where impulse purchases are concerned because it attracts the attention of the customer and may create the sale; it is also particularly important in supermarket shopping where few people have time for mature consideration and where the package that appeals to the customer is the one that sells.

The whole purpose of the graphic design of a package is to attract the public at large and colour is a vital part of this process. The problem is to decide how colour can be used to best advantage to achieve the aim, and it is worth considering what colour can do for a package:

- The first objective of a package is that it should command the eye, and colour achieves this.
- The successful package commands attention and triggers action whether the purchaser likes it or not; the physiological effects of colour help to ensure maximum attention value.
- The package should have maximum visibility and recognition qualities; the psychological effects of colour will ensure that people identify the package when it is displayed.
- The package should influence people to look at it closely and purchase it; colour will help to ensure that the package sells.
- The package should draw attention to itself and hold that attention; colour helps to achieve this.
- Colour is the vital influence that completes the sale.
- Colour can make the print easier to read.
- Colour helps to support the sales theme.
- Colour helps to co-ordinate the package with other forms of promotion, particularly television.

It is equally difficult to decide which colours will have most effect on the customer at point of sale. The colours must be selected with great care and this is a matter not only of hue but also of variation of hue. Colour selection is not an exact science, nor is it possible to ascertain the right colours by asking people what they would like. Colour must be selected to support a specific marketing objective or sales theme and in some cases may be selected for purely functional reasons, for example to protect the contents from light.

3.5 Trends

There are trends of preference for design and trends of consumer preference for colour, both of which must be taken into account when designing a visually attractive package. The design should reflect current thinking on the part of the average purchaser and it should not date too quickly; a design which seems to be out of date can lose sales. However, this is not to say that design should be changed frequently; too frequent changes are to be avoided because they run the risk of losing the image of the brand or product.

It has been mentioned in Section 3.3 of this part that the majority of people are attracted by smooth, flowing lines and that these are a trend of the times; an older package design can often be modernised by softening the line. Design trends do not alter overnight but ideas do change over the course of time and the skilful designer will keep the situation under review.

Trends of consumer preference for colour are even more important than design trends. Packaging is significant primarily in the domain of consumer products and therefore human emotions and preferences are vital. Vogues for colours rise and fall and the vogues should be reflected in the colours used for packages. The housewife will expect a change from time to time and, although it may not be practicable or wise to change the basic hue, it can be brought up to date by changing the variation to reflect current preferences. It is hardly to be expected that any consumer will get excited over a colour that was selected ten years ago. Whatever is being bought, the people who have to be attracted are the same and their wishes must be gratified if the package is to have maximum visual appeal.

The package design does not only have to reflect current preferences. Future trends have to be considered to some extent because the package is likely to remain on the market for some considerable time, depending on how easily it can be updated and whether there is a policy of periodical change. This needs careful planning and a study of current and future trends of consumer preference for colour in the home. If people like a colour sufficiently well to purchase it for use in the home, they will also like it when they see it on a package on the shelf.

A hue, or variation of hue, which is popular by current standards will make the package more likely to be noticed, picked up and studied, but this does not mean that trends should override all other considerations. First select colours which have practical marketing benefits and then adjust the hue to reflect trends. Trends should take second place to physiological and psychological factors.

Trends may be more important in some cases than others. Textiles and hardware are specific instances, and attention should be directed to trends when the package is likely to be used in the home, for example in the kitchen or bathroom. The fact that a package does not look out of place in a modern kitchen may be a subtle selling point. Fashion trends in clothing may be significant in some instances and there may be changes in preferences for colour in relation to specific products, dictated by natural progression or a desire for change. These should be watched.

4 Ideal design

4.1 The approach

It will be obvious from the previous sections that a successful package has to be designed, and the term *package design* means different things to different people. To some it means functional or technical design, that is the way that the package is put together; to others it means graphic design, that is the picture on the pack. In fact, package design is a total concept which embraces functional design, perception (shape, form and colour) and graphic design, all of which should co-ordinate to produce the objectives that the package has to achieve as part of the marketing process.

A printed package is arguably the most important material issued by any company and the whole justification for the design process is that it helps the company to make increased sales and larger profits. If the package is to achieve its objective, the design must be a compromise between a number of conflicting interests; all the various features which go to make up the ideal package must be related to each other and combined into a whole which has a specific target.

Package design is concerned with the way that a product sells, but it is also concerned with the economics of packaging and with the production line. The design must be an interpretation of the marketing plan which will communicate with all the parties involved, thus selling the product and making profits. A package is a means of communication rather than an art form but its design is also concerned with costs; valuable savings in production costs will often devolve from small changes in design and, for example, a dual-purpose package which also helps to display the product can halve the total cost of packaging.

Package design is ultimately the responsibility of management and, although it may employ the creative talents of a professional designer, management must specify the target and provide the designer with adequate data on which to work. Management must decide what it wants, define the market at which it is aiming, discover the needs of the retailer and consumer and co-operate with the designer to formulate the most practical and economical form of packaging.

Although design is the ultimate responsibility of management, it is arguable who should undertake the actual work. There are a number of options available:

The specialist designer   Independent, more objective and a specialist. In the US every company of standing uses a specialist.

Advertising agents   Supposed to be nearer to manufacturers and consumers and can design with advertising in mind. This may be important but many agencies lack the specialist skills required, especially the functional aspects.

Staff designers   Nearer to the problem but may not be able to see the wood for the trees and cannot be as objective as might be desirable.

The package manufacturer   Many complain that graphic designers neglect the technicalities and that artwork does not fit into machinery specifications. May be liable to emphasise technical aspects at the expense of marketing.

It is no part of the function of this guide to recommend one option or the other, but experience suggests that there are many arguments in favour of the specialist designer. The role of packaging in marketing is crucial and specialist designers have shown that they understand the business of selling and most have a businesslike approach. A well known American company operating in this country pointed out that package design can make or break a product, and they believe that the placing of the different elements of the design is an integral part of the whole; this requires considerable study and experimentation and this causes them to seek expert help from outside specialists.

The primary task of design is to communicate, through the shape of the package and its graphics, a specific, unique and believable image. Package design is a statement of product individuality, a stimulus to purchase and a motivation for brand loyalty. The designer must know something about the desires and feelings of the average customer and must resist the temptation to impose a personal 'mark' on the package without taking account of all the marketing considerations involved; that way product personality may be lost. The designer may have to sacrifice design themes near to his heart on some occasions because the client will not 'buy' them, but all parties should recollect that the object of design is to *sell the product*.

Package design seldom requires a high art form and too much subtlety may defeat its own object; there is a tendency to complicate what is really a straightforward problem. Many designers have the idea that it is their function to mould and guide the tastes of the public and they resent the idea that design is concerned with anything as mundane as sales. This does not apply, of course, to the skilled specialist.

A bowl of sugar or a pile of peas are not merchandising elements in themselves, but through the medium of package design they become dramatically individual and set up a whole profile of attributes and impressions in the minds of consumers, thus triggering off the desire to buy. The package design must satisfy the need for protection of the product and all the other factors that go to make the ideal package and must convey the essence of the product in the most psychologically effective manner. It must stir the heart as well as catching the eye. An ideal design must appeal to the emotions as well as to the mind and should do everything possible to convince the purchaser that the supplier has the best interests of the consumer at heart. It should give pleasure to handle and be seen.

The ideal design depends on the creative skill of the designer, provided that the designer is not restricted by so-called factual analysis. Very often creative skill is restricted by fact finders who are not sure what they are trying to find, or what facts they ought to collect. Too often the creative approach is confused by academic and arbitrary testing methods which mean very little. Nothing in this book is intended to suggest any restrictions on the creative skill of the designer; it is necessary to take into account various marketing and functional characteristics, but the psychology of the consumer is far more important than mechanical measurements.

4.2   Research for design

Despite the strictures on uninformed fact finding at the end of the previous section, it is necessary to undertake adequate research at the outset of the design process. If a package is to be effective, all its elements must be chosen for their marketing effectiveness and the designer cannot know which shape, size, colour, material or illustration will communicate the right message without adequate information. Research is the tool which provides this information and which pinpoints the objective.

Good designers are aware of the need for research and many carry out their own, but some fear it and are afraid that the results of research will replace creativity or even make the designer obsolete. This, of course, is nonsense. The purpose of research is to provide a comprehensive brief for the designer encompassing all the factors that may affect the end result. These factors include product characteristics, the market at which the product is aimed, the competitive situation and other relevant information. Only adequate research can bring to light all the marketing considerations which will create product personality. The well known American designer, Walter Lander, describing the extensive tests and research that he carried out before, during and after every assignment, explained that in his view the designer must not put his personal signature on the design that he creates, nor must any style similarity be allowed to creep into the work. The client is entitled to expect a unique personality for his product, reflecting the circumstances specific to his brand, his product and his marketing situation, and this unique personality can only be achieved if the potential market is properly researched.

It is necessary to study all aspects of the situation and to relate the package to the product, to the company, to the market concerned and to the brand or corporate image. This review should bring together all possible knowledge about any factors that might affect the ultimate result; this may require market research, but will certainly require study of market requirements as well as gathering together any data which may affect the colour and shape of the package. Research should cover all possible contingencies so that management has adequate information to ensure that the package does its job properly. The research may include the following:

Internal research   Research within the organisation is necessary to discover and evaluate the interrelationships between products, printed matter, facilities and display. It will also seek to initiate and implement co-ordination between departments to establish the objectives to be achieved and to interpret those objectives so that they can be translated into visual terms. The study should embrace the corporate image and whether this should be reflected in the packaging; it may even be necessary to define the corporate image and corporate strategies that management may wish to be reflected in packaging. Policy may be to build up an individual brand image.

Market research   This is necessary to ensure that the package is aimed at the right market, for example

- how people buy the product in question
- how people react to various stimuli
- the type of people who buy, and whether young or old
- the selling conditions that are usual.

The package must play an integral part in the marketing plan. The purpose of the market research is to establish a valid picture of the market; it is not to ask people what they would like.

Technical research   Research should establish that the package will protect its contents adequately; that it will fit into the distribution network; and that it will be suitable for the plant and machinery available. Study of the processes of production and channels of distribution for the product in question are necessary, together with a profile of the most convenient forms of packaging. It may be useful to study alternative materials and alternative types of package, in the interests of cost and efficiency. Newer, and more recently developed, forms of packaging are worth looking at; they may have marketing advantages. Costs also have to be considered; costs of retooling may be too heavy. Most technical research will involve study and discussion of the various factors involved, plus application of the experience of knowledgeable people. It may be useful to take the opinions of wholesalers and retailers, particularly on such matters as the value of existing packs, ease of handling, ease of pricing and similar matters.

Competition   A study of competition is necessary to make sure that all concerned are aware of the activities of competitors and the nature of the packages offered. Any package should be different from, and better than, those offered by the competition. Study should also include those packages that are likely to be displayed in close proximity to the package being evaluated. A collection of competing packs for comparison purposes will be useful.

Economic research   This may include study of the supply and price structure of raw materials; the cost effectiveness of different types of package; the desirability of holding stocks of raw materials or finished packages; the capital value of new machines and their possible effect on production costs; and other matters of a similar nature.

Consumer research   There is little point in asking consumers what they would like, because they are not aware of their own subconscious reactions and their opinions will be governed by anxious thoughts which will not reflect their true behaviour. The opinions of consumers about the appeal of a package are largely valueless and it is better to make use of established psychological and physiological principles. A good package has to appeal to the *average* consumer and not to the individual, and there is no place for personal preferences.

Much research into packaging is of a very arbitrary nature; studies concerned with semantic differentials, controlled associations and the like are largely academic in nature and the researcher frequently credits people with responses which are wholly superficial. Many of the so-called scientific tests are downright misleading. In recent years some companies have tried researching package design by means of group discussions, but these can be misleading because they do not reflect actions at point of sale; in such a discussion people concentrate only on the package, whereas at point of sale they concentrate on the product. At point of sale the package *is* the product and there must be a close association between package, product and market. Research is not creative; it produces not a finished design but facts and figures which have to be translated into a package via a brief for the designer.

Retail research   Research may be required at retail level to discover the needs of retailers, supermarkets and the like. This is normally part of the technical research mentioned above, but special attention may be necessary to changes in retail trade such as the incidence of pallets, shrink wrapping and the like.

Trends   Research is also necessary into trends of consumer preference for design and colour. This is discussed in the previous section.

A word of caution is necessary. There is a risk that market research can be taken too seriously and that this may deter the manufacturer from being too entrepreneurial. It will not, for example, tell the manufacturer whether a new shape is required or what colours should be used.

4.3   The brief

The object of the research mentioned in the previous section is to prepare a brief for the designer, but the process can often be simplified and a great deal of unnecessary cost avoided if packaging is considered at the design stage of the *product*. For example, a 75 mm knob might require very complicated packaging and the cost of the extra packaging might be out of all proportion to the value of the knob. However, at whatever stage the design of the package is considered, a great deal of data is required to pinpoint the market, identify the customer and discover the requirements and preferences of the customer. All this has to be incorporated into a coherent brief; this is often more difficult than it sounds because there will be a number of people concerned with any packaging problem, and each of them will have their own ideas about the correct solution of the problem.

The important point is that the design brief for any package should take account of all the factors concerned. An excellent graphic design may not be appropriate for the type of package which is most economical and which best suits the production line. On the other hand, a well thought out concept may be spoilt because the buying department tries to do the job on the cheap. There are those who hold the view that a package should mirror the prejudices of the managing director because he has the ultimate responsibility for the image of the company, but few managing directors today would want to take that responsibility without expert advice and assistance.

Someone has to take the responsibility for the preparation of the brief, has to set it down, has to think about each facet of the package, has to apply common sense and has to have more than a passing knowledge of packaging materials. It is useful to know, for example, which form of package

- fills fastest
- uses the least expensive materials
- requires the latest complex machinery
- stacks well in store and does not create inventory problems
- is robust enough to stand up to the product
- will best adapt to labelling requirements appropriate to the product.

The process of brief preparation involves reviewing the whole situation, co-ordinating all the various factors involved, ensuring that no relevant points are overlooked, and collecting together data about all aspects of the situation. The person concerned has to have a pretty agile mind which is alert to changes in packaging techniques, retail environment and graphic styles. Much of the data will be the results of management decisions about marketing policy and these decisions have to be taken before any design can take place. All of this presupposes that a marketing plan exists and that the nature of the package can be decided before anyone thinks about visual and graphic display.

The package, and package design, is an essential part of marketing, and the way that the package is to play a part in the marketing plan is fundamental to both functional design and graphic design. The preparation of the brief is therefore an essential part of marketing planning; the elements of the market plan dictate the whole approach. However, what may be ideal from a marketing point of view may not be practical in functional terms, and this may require a good deal of work and research before suitable solutions are evolved which are agreed and understood by all the parties involved. The eventual solution should be practical and as simple as possible; a complicated package costs time and money and will inevitably cause production problems.

A typical example of the process in action is a package for an Easter egg. This has a fairly obvious aim and, in the first place, the package must be designed to protect the egg. It will also have to have a graphic design appropriate to the season and to the ultimate purchaser, and this depends on the definition of the customer at whom the product is aimed.

It may be aimed at the child who is the likely consumer or it may be aimed at the adult who gives the egg as a present. Because the graphic design includes a chicken, it is not necessarily appropriate to both child and adult. Adults will buy what *they* think will appeal to the child but the latter may have quite different ideas. It may be necessary to study the market to decide which is the most profitable sector and to find out what considerations are likely to govern the choice of the purchasers.

5 Changing packaging

5.1 Policy for change

However much care is devoted to the design of a package, there comes a time when it needs to be changed. Wise management will keep packaging under constant review, making a point of regular consideration of the need for change. Ideally, a new package should start being planned as soon as the previous one goes into production; in that way a new approach is ready when circumstances demand it. Those circumstances may be one of a number of reasons, including falling sales. Although a flattening of the sales graph may not be due to packaging, it is worth looking at.

Packaging may not require revolution but it may benefit from evolution. No package design lasts for ever and the best selling package benefits from gradual change which will keep it up to date. A package which once seemed fresh and attractive may have become dated and, while a slightly old fashioned look may be no bad thing, a senile image can kill sales. It has been said that old brand leaders never die, they simply require repackaging, but this can be a very delicate operation because any impression that the product has changed must be avoided (except, of course, where the product *has* been changed or reformulated and a new package will help to create a new image). Drastic surgery may alienate brand loyalty and give consumers the idea that a changed package indicates a changed product. Where a new approach may be desirable for one reason or another, a change that is too radical may endanger the brand image altogether.

There is often a marked unwillingness to make changes in packaging because the package has become an integral part of a brand or corporate image. There may be opposition from sales people and retailers to any change which might interfere with things as they are, but experience suggests that innovation rarely harms sales provided that the new package retains some of the elements of the old.

When considering change, each detail of an existing package should be taken apart and considered in relation to the image of the product. It is particularly important to identify the key recognition factors so that modifications will not do away with them or alter them. Research will seldom indicate what should be the end result of change and a great deal depends on the skill of the designer.

5.2   Reasons for change

There may be a number of reasons for changing the design of a package, for example because of

- falling sales
- superior packaging by the competition
- changing consumer trends
- changing consumer attitudes
- changing market conditions
- new marketing policy
- new developments in packaging materials
- new retail developments.

It is surprising how few managements consider packaging when sales begin to falter. The package may have lost its pristine attraction; the design may be dated; the type of packaging may have become unacceptable; or the package may no longer do its job properly because of poor materials or for some other reason. There may be a demand for something new, and a new approach can revive the fortunes of an also-ran by improving its image. A change in packaging is almost always good for sales whether or not it is accompanied by a change in sales strategy, but it need not be a radical change. A new package may attract new customers but some existing features left as they are will help to retain the loyalty of some old ones.

Competition between most packaged products is fierce, and this dictates constant review of packaging concepts and packaging costs to ensure that the product remains competitive and compares well with the competition. The competitive situation might be such that a complete rethink is required; economies in packaging might make it possible to reduce selling prices, thus creating competitive advantage.

Consumer trends may suggest change because people at large are affected by developing trends in architecture, in art and in colour, and these may affect the acceptability of graphic design. Today's designs may be considered old fashioned tomorrow. Consumer attitudes to a product, and the way that it is presented, may change after a time, perhaps because of changing life styles and perhaps because of changing social conditions; pressure from consumer organisations and care for the environment often predispose people against a form of packaging. Changes in the way that a product is used may necessitate major changes in packaging.

Changes in retail selling techniques often dictate changes in packaging methods. Supermarkets changed the whole philosophy of packaging; shelf appeal became more important that protection. The quality and standard of the package may have to be improved to meet new conditions, and 'once a week' shopping may require different sizes. New developments in retailing are of many kinds; supermarkets now stock far more than groceries and these developments may require a rethink of packaging.

Products have lives, just like people. When a product reaches maturity something may be required to restore youth, and this something may be a new approach to packaging, a completely new image, a new slogan or even a new marketing plan. All require a rethink about packaging. The packaging industry is a dynamic one. Every new development, every new material and every new type of package ought to be examined to assess its possibilities and whether these can make a contribution to better marketing. A new development may be just what is required to add a little 'plus' to slow sales. A study of available packaging materials will often bring to light something 'different' that will create a new individuality, but it is also useful to consider other ideas – a new functional concept, a different size or a different type of packing.

5.3   Design factors

Special considerations apply when change of packaging is contemplated, and the designer needs to know why change is being made and what it is desired to achieve by making changes. Any change should be based on good evidence and not on guesswork, and all possible care must be taken not to disturb an existing image.

Package designs often 'grow up' over the years and become visually cluttered because no one has taken the trouble to review them. It is seldom a good idea to redesign totally, although there may be a case for doing so in some instances (and this does not mean to say that the whole situation should not be *reviewed* from the beginning). Where there is an established product image, any new package should remind people of the old and a link between the two is advisable.

Consider the following points:

- Possible consumer attitudes to change. When Bird's Eye first put peas into bags, customers thought that they were getting better value from lower packaging costs.
- Pattern of usage in the home. This may be changing, and changes in the nature of the package may be useful, for example larger, smaller or more convenient.
- Where the product is displayed on the supermarket shelf. Changes in retail practice may suggest modifications to the package or to its design.
- Changes in competitive packages, or a new sales theme from the competition.
- Whether the cost of change will be justified by increased sales. Change for the sake of change is seldom useful.
- Whether radical redesign is required. A change of typeface or a change of colour may be all that is necessary; old fashioned typefaces and out of date colours can create a wrong impression.
- Do not change the design because of an advertising campaign. The campaign may be short lived, but make sure the package design reflects the overall sales theme.
- Graphics must change with the times. Update illustrations as well as style, but the design should not date too quickly.
- A strong package may no longer be necessary; but better protection may be.

6 Packaging assessment

6.1 Assessment

Assessment means studying a package, and packaging generally, from all points of view in order to provide management with data and recommendations. It is often useful to have an assessment made by someone who is quite independent of a company, who has not absorbed existing ideas and is not constrained by company policy; it is particularly useful when new ideas and a fresh mind are required. An assessment may be made either when a new package is being planned or when revision of existing packaging is being contemplated. In either case it will be desirable to review the whole packaging picture and the whole field of related packs in order to maintain continuity of image.

There may be a number of reasons for making an independent assessment, for example

- to provide a brief for the designer
- to provide management with a blueprint for packaging policy
- to give management fresh ideas for a new package
- to resolve conflicting interests on a committee
- to bring a fresh mind to bear on problems
- to provide a check on company thinking
- to review existing packaging in order to decide whether it would benefit from change
- to ensure that all relevant factors have been taken into account in planning
- to make sure that packaging is contributing to the image of the product, and the company
- to check that packaging is cost effective
- to check that packaging is being carried out in the most economical way

40

- to ensure that packaging is making a maximum contribution to sales.

The important point is that the situation should be studied from all angles – marketing, functional and graphical – and the study may involve market research, research into the technical aspects of packaging, research into the availability of materials, and the application of expert knowledge of shape, form and colour. Such a study is not creative, it is purely factual, and for that reason is best carried out by someone with experience of packaging, but not the designer. Although assessment is part of the process of creating individuality, and does not replace the creative skill of the designer, it does require something more than creative skill.

By investigating a situation from every angle, an assessment helps to ensure that all alternatives are considered including new raw materials, new functional concepts, changes of style or size, and possible modifications making for easier handling in distribution.

As assessment might be concerned only with, say, improving colour but it must take into account all the factors that might affect colour, including the marketing picture. Any package will fail in its primary objective if there is no clear conception of the part that it might play in the marketing plan. It follows that an assessment which starts at the beginning is most valuable.

The stages of an assessment might take the following form:

- examination of the whole packaging process from production, through distribution, marketing and merchandising, to the ultimate consumer
- study of alternative materials and types of packaging, and the economics of packaging
- research into the marketing situation, including the task that the package has to perform
- study of the merchandising aspects, wholesale and retail aspects, and selling conditions
- study of the competition and its packages, including any packages which are likely to be displayed in close association even though they are not strictly competitive
- assessment of the most suitable type and form of packaging to correlate with the marketing and production needs
- analysis of what the package has to do and how it can best meet the needs of the market place
- analysis of the most suitable shape and size of package
- evaluation of the essential features of design necessary to meet the marketing target
- recommendations about the most suitable form and shape of the graphic design

- recommendations about the most suitable colour or colours for the package
- preparation of a brief for the designer.

6.2 Evaluation

When the design of a package has been finalised, management often feels that it ought to have further tests of the acceptability of the package to the consumer, but this may be a waste of time and money. A great deal of skilled creative talent is wasted because of arbitrary testing methods and academic ocular measurements which do not prove anything at all. It is virtually impossible to duplicate point of sale conditions by artificial means and this applies particularly where colour is concerned.

This report has sought to point out that acceptability of a package is very largely dependent on emotional and psychological factors. These cannot be tested by normal market research methods because opinions expressed in a poll do not reflect actions taken at point of sale, and mechanical testing will invariably show the best results where contrasts are strong, irrespective of the emotional acceptability of the package.

The weakness of mechanical testing can best be illustrated by considering the appearance of colours. The red of a glass of wine, the red of a piece of silk, or the red of transparent Cellophane are all *red*. They can be matched so that they have exactly the same instrument readings but they certainly do not *look* the same. It is the perception of the total concept that governs acceptance.

There is a good case for testing a proposed design under different conditions of lighting and in actual display conditions to ensure that the design and the colours are not overshadowed by the competition or become 'lost' in certain lighting conditions, but opinions would be subjective rather than objective. There is also a good case for exhibiting similar, or near similar, designs in different shades to consumers to discover whether they prefer stronger or lighter colours; or, say, which of two different shades of green they prefer. In cases of doubt there may be a case for testing a specific concept, for example whether people prefer coffee out of a red or out of a brown tin.

From time to time new research methods become fashionable. Semantic differential tests were popular at one time and the tachistoscope was much used to measure the exposure required for visual recognition, but the writer is not convinced that these methods have any real value. Far better to rely on established rules of perception and to test specific concepts when necessary. The great thing is to apply common sense. By all means use market research, and do not depend on individual prejudices, but make sure that the market research will provide valid results.

The following are some comments on the more usual methods of testing:

The eye observation camera   This works on the principle that the pupil of the eye, besides adjusting itself to the amount of light reaching the retina, will also dilate or contract according to emotional involvement in the object that is observed. Adherents of this camera claim that use can provide a more accurate prediction of market behaviour than verbally expressed preferences, and it is less costly than in-store tests, but should be supplemented by other methods of testing. The fallacy is that the pupil dilates according to the amount of light, and this proves nothing – certainly not the emotional claim. The emotions aroused may have nothing to do with the object being viewed, or with whether the package will sell or not.

The tachistoscope   This device exposes individuals to the package for short but increasing intervals, and after each exposure they are asked what they have seen. Adherents claim that this indicates how the package rates on impact and recognition compared with its competitors and what impression is conveyed by the product line. The fallacy is that black and white illustrations will always show the best results because there is maximum contrast between the two colours. Also, the eye may concentrate on some bewildering or puzzling point and thus convey entirely wrong results.

The anglemeter   This measures the angle at which the package is recognisable on display. It is claimed to be important in supermarket selling but, in fact, no two conditions of display will ever be the same.

In-home tests   These are designed to show how well the package will do its job as a container in the home. Such tests may be useful in some cases.

Spectrophotometric measurements   This instrument simply measures and analyses colours. It is an essential tool of the technician who has to maintain colour standards but it has little application from a package testing point of view. The colour memory of the average individual is notoriously poor.

Hall tests   Packages are exhibited to individuals who are then questioned about their reactions. These tests are useful in solving specific problems, provided that the questions are carefully thought out and that they are the right questions to ask. It must always be remembered that opinions expressed today do not necessarily reflect action at point of sale tomorrow, and that where colour is concerned emotion is much more influential than any visual aspect.

Every research expert will have his or her own ideas on the most suitable methods of testing packages, but in the long run the most satisfactory solution to the problem is to use common sense in applying well established principles to carefully thought out data.

Part II
GRAPHICAL APPLICATIONS OF COLOUR

1 Explanatory

This part is concerned with the application of colour to graphic design in the production of promotional material; it is concerned with broad principles rather than with specific situations. The application of the principles to individual circumstances and to specific types of packaging is discussed at a later stage. The text that follows suggests ways in which colour can be used to maximum effect, outlines the broad principles involved in finding the right hue and provides an introduction to the selection process. It is not concerned with environmental colour, with the choice of colour for products or with the technicalities of inks or pigments, but it is concerned to show how the right combination of colour and graphic design can produce effective results.

The intention is to provide a link between marketing and design but not to pre-empt the creative skill of the artist or designer. The ideas expressed may be of assistance to designers but are primarily intended for management and for those who are responsible for the initiation and production of promotional material in general, and packaging in particular.

The main emphasis is on promotion aimed at the ordinary consumer but the same basic principles apply to material aimed at industrial targets, although the considerations governing choice of colour may be different. Although this is not a treatise on the finer points of graphic design, and is not concerned with the technical aspects of production, both are mentioned briefly where they are relevant to the principal aim of using colour to the maximum benefit of sales. Selection of the right colour, whether for packaging or any other promotional material, is an essential part of marketing, and from this it follows that the first step in colour selection is to define the marketing target quite clearly.

Packaging is, of course, embraced within any definition of promotional material and graphic design is an essential element of the ideal package. Therefore the broad principles governing the application of colour to the graphic design of promotional material apply with equal weight to packaging and, indeed, should be clearly understood because of the desirability of co-ordinating package design with other forms of promotion.

If any promotional material is to be effective it has to achieve three primary objectives:

- It must attract attention.
- It must create a desire to read, or take notice.
- It must communicate a message in the most appealing way and trigger action where necessary.

To achieve these objectives the promotional material must have visual appeal and packaging must also have visual appeal, however important its functional and practical aspects may be. This visual appeal is a combination of the two elements mentioned above, namely:

Graphic design, including shape, form and balance. In the case of packaging, the shape and form of the package itself is also significant.

The colour, or·combination of colours, used as part of the graphic design or as a background to it.

The variables that influence results require careful and thoughtful consideration, and just as much care and thought should be given to effective colour as to good layout and typography.

2　About design

2.1　Graphic design

The first of the two principal elements which go to make up visual appeal is graphic design. Although this book is not intended to be a treatise on graphic design, it is desirable to say a few words about it because design and colour are closely allied.

It has been said that good design is the most effective factor in any promotion, whatever the medium, because it goes on working long after the original promotion has been terminated. This is because good design has an effect which is largely subconscious. People do not generally realise that they are influenced by design and they do not study, and analyse, each element. Graphic design includes illustration, text, typeface, layout and – where appropriate – colour, all combined together to create a total impression which is absorbed subconsciously. Even the environment in which the image is seen may well alter the interpretation created in the mind of the viewer.

This is no place to discuss the merits of various schools of design, or the advantages or disadvantages of using outside designers or staff designers, but it can be said with certainty that there is a need for the creative skills of the experienced designer, if only to avoid the sameness of much promotional material. There can be a suffocating similarity of typography and layout in all promotional material, and particularly in packaging; this is a serious fault. On the other hand, there is also the risk that the creative designer will express too much of his own personality and produce a result which is way over the heads of the people it is supposed to attract. If the designer is to earn his keep, the result of his labours might just as well appeal to those who see it and therefore do something to aid sales.

Some designers have the idea that it is their function in life to mould and guide the tastes of the public and they resent the suggestion that design is concerned with anything so mundane as sales. There is a tendency to complicate what is really a straightforward problem and to forget that artistic designs do not necessarily have marketing appeal. Those who commission or buy graphic design are concerned to attract the attention of the public at large and, having done so, to impel that public to take action. All parties concerned are, or should be, vitally interested in ensuring that creative skill is used to make design effective as a selling proposition.

In fairness to designers it must also be said that there are often corporate barriers to creativity. Some authorities aver that British manufacturers tend to dictate to designers too much and the same manufacturers often regard colour as a dirty word. EEC competition has had the beneficial effect of persuading manufacturers to give more attention to good design.

There can be much argument over what constitutes good design. However, it is certain that to create the visual appeal which is so important for promotional material, there must be an understanding of the principles of human perception and they must be put to good use. Evaluation of any promotional message by an individual goes far beyond the eye; it involves the brain and the human psyche itself. What the brain does with the message it receives is particularly important. The work of the psychologist has a direct and practical bearing on the subject because it identifies the whys and wherefores of human action and behaviour.

The following are a few broad principles suggested as helpful in achieving effective design for promotional material:

- The impression should be pleasing from both distance and short view.
- The design must tell the sales story quickly, easily and logically.
- Always reflect fashion trends in design and colour.
- The eye prefers to move horizontally; vertical movements are boring.
- Clear vision requires the eye to be fixed; it is temporarily blinded when in motion.
- Information should be presented logically.
- A well ordered layout suggests restraint, dignity and self-assured salesmanship.
- A haphazard layout may attract attention but lose a sale because the customer does not bother to read it.
- Certain forms and shapes become distorted in the angles of peripheral vision.
- A distorted shape will do more harm than good.

- An illustration that is too dominant may attract attention away from other important features.
- If motion is implied, move the design off balance, for example rotate a square so that it becomes a diamond; this is more dynamic, provided the shape is regular.
- Convexity is more pleasing than concavity. Convex forms convey an impression of billowing and expanding; concave forms tend to collapse.
- Simple geometric forms have more impact and memorability than irregular ones.
- Regularly spaced intervals are stable and satisfactory; haphazard intervals are objectionable.
- A horizontal line based on an imaginary ground conveys tranquillity; a vertical line implies activity.
- Form and shape should be tactile and soft.
- Contrasts of light and shade should not be too sharp.
- To trade up, use a more sophisticated design.

2.2 Perception

Perception can be defined as the study of why human beings see as they do and how they react to shape, form, balance and colour. All of these are important in the graphic design of promotional material, whether it be press advertising, direct mail shots or the label on a can. Perception has nothing to do with subliminal advertising or any other 'funny' business, but is simply a combination of fundamental physiological and psychological principles in which optical considerations, emotion and reasoning all play a part.

A design makes an impression on the mind going far beyond what the camera records, and effective design must take this into account along with all the other factors involved, including functional considerations in the case of a package. Visual appeal requires the application of principles of *Gestalt* psychology; if the piece is to have maximum appeal it must make an instantaneous impression which is simple and direct because exposure to the eye is usually of short duration. Simplicity and regularity are vital.

However, instantaneous impressions are of little value if the design makes no lasting impact. It is here that colour is so important because it helps to create impact, increase appeal and provide better recognition quality. Many of the principles of perception are common sense and, although they have a scientific backing, there is no need to apply them in a scientific way. The aim is to ensure that a design attracts attention without conscious effort on the part of the viewer. Much graphic design is neurotic because it tries too hard and makes a spectacle of itself in an endeavour to attract attention. Simplicity is much more effective.

A first principle of perception is that the eye and the brain demand simplicity and balance in everything that they see; if these two do not exist, the eye and the brain will try to create them. This principle influences the form of a design and the shape of a package and helps to create memorability. Attempts to be too clever will defeat the object of an exercise and are unlikely to achieve a useful result. If an irregular form is seen on short exposure, as it is with most promotional material, the mind of the viewer will tend to smooth out sharp angles and to see the form as regular because the eye and the brain want it that way. Anything that is askew bothers the eye and creates a mental block, and this causes impact to be lost. Where there is too much complexity, viewers tend to close their eyes (metaphorically speaking) and to reject something that is too difficult to understand. The majority of people are more favourably influenced by smooth, flowing lines than by hard, angular ones. No design should ever be so 'hard' that it repels people; nor should it present sharp angles or be 'cut in two', perhaps by a brand name. Although design trends do change, it is safe to say that simplicity is always best for graphic design of a promotional nature; form, proportion and balance are always important.

Another basic principle of perception is that in any design the eye persists in isolating the main part, usually the centre, from its surround. The isolated part will be given solidity and detail while the surround tends to be seen as soft and yielding. Vision always tends to concentrate its attention on the principal part, and this is where emphasis should be placed. Undue distraction from the principal part will tend to bother the eye of the viewer and lose the impact of the design; in the case of packaging it may deter purchase. A simple design will help to concentrate the attention of the viewer and is strongly recommended for packages bought on impulse.

The most satisfactory way of achieving maximum concentration is to use a symbol on which the eye of the viewer can fasten; provided that the symbol is easily recognisable, its use will ensure instant recognition. The symbol can be a name, a trade mark or just an abstract shape. This strategy is particularly useful in packaging design but can be used for other material as well.

Any design which is a jumble of conflicting shapes and forms will be 'lost'. The recommended strategy is to ensure that the attention of the viewer is concentrated on the symbol, and then to use the natural habit of most people to read from left to right. If, therefore, the symbol is placed slightly to the left of centre of the design, the eye of the viewer will tend to concentrate naturally on it and then to move to the secondary item which should be placed slightly to the right; the secondary item might be product description, price or some other feature. Good colour will help to underline the strategy; both symbol and secondary item should be in stronger colours than the background, which should be kept soft. This will achieve maximum impact. Some modification of this principle may be necessary in those cases where the normal reading habit is from right to left, for example Arabic.

3   About colour

3.1   Colour in promotion

Colour is the second of the two elements which make up visual appeal, and its importance lies in the fact that colour appeals to the emotions rather than to reason. People like colour and they react to it at a subconscious level; an innate liking for colour is part of the human psyche. It helps to reduce sales resistance and ensures that graphic design has maximum appeal; it is a vital factor in creating graphic designs that sell. Colour achieves its target by reason of its visual and mental attributes:

Physiological responses   Colour catches the eye and invites attention, however colourless the message conveyed. It may create a sale by impulse attraction at point of sale.

Psychological responses   Colour can help to express warmth, coolness, quality, moods and other emotions because colour is based on human nature. It can convey an impression of the seasons, such as spring and summer, and other desirable feelings.

The appeal to the senses   Colour can add dimension and realism to products whose appearance cannot readily be conveyed without it.

The appeal to the emotions   Colour may be used for the pleasure that it conveys and to improve appearance; however, the user may need to be reassured that this will increase sales.

The basic reason for using colour in any promotional application is to attract attention and thus help to sell the product, service or concept concerned. The best colours for this purpose are basic, simple colours which reflect physiological responses. When the eye looks at the spectrum it does not see an infinity of colours but sorts them into groups, and these elemental groups are hard to surpass in any graphical applications. When colour areas of the same brightness are seen side by side, the warm colours – red, yellow, orange – will segregate better than the cool colours – blue, green and violet. The warm colours will be sharp and clear cut and for that reason are also called hard colours; the soft colours will tend to merge together and are therefore called soft colours.

Of all the colours, red, yellow and orange are the best attention getters and cannot be disregarded in any circumstances whether or not a person likes them emotionally or aesthetically. Elements in a design which are meant to stand out should ideally be in hard colours and preferably set off against a background of soft colours, but the ideal may be modified for psychological, emotional or marketing reasons. Colour can do more than just attract attention and it is the psychological and emotional responses which provide the icing on the cake, so to speak.

Colour has been described as a unique form of communication which cannot be compared with other methods. Whether one is selling, telling a story through advertisement or communicating in a scientific sense, colour adds dimension to communication. Among other things, colour

- activates
- stimulates greater identification
- ensures emotional participation
- adds mood
- creates an internal cohesion in the message
- is perceived before form
- is perceived earlier in the life of an individual
- is immediate and emotional: words have to stand for something
- does not have to be translated: it can be understood directly
- adds permanence.

In some cases colour may actually create a sale by catching the eye and impelling purchase. Supermarkets are a riot of colour because colour is considered to be essential to moving merchandise. At point of sale, colours of displays and background help products to battle it out in the most competitive form of selling, and this applies equally to packaging; breakfast cereals in a black and white box just would not sell.

Although it is hard to beat the basic colours for attracting attention, they may be combined with more subtle shades to create a specific emotional appeal; subtly contrasting or unusual colours can communicate a message in addition to having basic attraction. The use of modified colours is not generally recommended in graphical applications but they may be used when something out of the ordinary is required or, perhaps, in high fashion markets.

The art of using colour in any promotional application is to find out what will appeal to the public and how colour can be used to best advantage at point of sale. However, before pursuing this point it will be useful to summarise what colour can do.

3.2   What colour can do

Colour is one of the most potent tools in the selling process, but it must be the right colour. Sales figures prove that the right colour increases sales, but the wrong colour kills them. The first function of colour is to attract attention, but sales depend on what the brain does with the message and this depends very largely on the hue selected. There is no point in using colour unless it is used effectively, and with good reason. This applies equally to product illustration, to the background of point of sale material and to the colour make-up of packages and labels.

In the promotional context, colour creates demand primarily by attracting attention and by exciting desire because it provides pleasure in a way that has an effect on the subconscious. For example:

- Light, pale colours, especially green and yellow, suggest the spring and will turn people's thoughts to new clothes and new furnishings.
- Certain colours have appetite appeal and will make food promotion more attractive, thus increasing demand for food.

People notice colour much more quickly than they notice shape or form and this not only helps to attract attention but holds that attention; an instantaneous impression is of little value if it makes no lasting impact. Colour is accepted or disliked intuitively whereas appreciation of design requires conscious thought; people are attracted by colour irrespective of education or culture but appreciation of form and shape requires experience.

Some of the things that colour can do include the following:

- Attract attention. The first objective of any promotional design is that it should command the eye; colour achieves this.
- Pre-sell the product. As with appetite colours, mentioned above.
- Excite demand. As with spring colours, mentioned above.
- Create an image. The psychological appeal of colour helps to ensure maximum visibility and attraction.
- Illustrate the product. Some products cannot be conveyed accurately in black and white.
- Inform about a range of colours. Shows people they have a choice.
- Provide an association of ideas. Important in a fashion context.
- Provide identification. An attractive combination of colours will remain in the mind of the viewer.
- Provide emphasis. The psychological appeal of colour helps to ensure maximum attention value.
- Shriek at the customer. When necessary.
- Identify a brand or a company image. Especially with packaging.

- Ensure maximum readability. Colour can make print easier to read; few people shopping in a supermarket, or glancing through a journal, have time for mature consideration.
- Ensure recognition. Colour will influence people to look at a design more clearly and to be influenced by it; colour helps the design to sell.
- Ensure maximum visibility. By careful choice of hue.
- Persuade the recipient to read. Pleasing combinations of colour and good contrast make reading easier.
- Impel action. Colour has more impact than neutrality.

3.3   Colour attributes

The objectives outlined in the previous sections can be achieved by selecting hues with the appropriate characteristics. Each individual hue has a number of basic attributes or properties which have been discovered by research and experience; they are partly emotional in nature and partly derived from the physiological, psychological and optical responses triggered off by colour. Before they can be used in an individual case it is necessary to identify those attributes and properties which are appropriate to the product or service, its place in the market and the objective of the promotion.

The following is a list of the principal attributes of colour which are relevant to packaging applications:

Age   Certain colours appeal to the young and less so to older people. Use colours appropriate to the type of person aimed at.

Associations   Many colours have associations with the environment, with products, with parts of the home, with religion, with politics and with other things. Some are traditional, some are common sense, some are practical. Select colours for packaging that are associated with the product, with the part of the home or with the specific theme that it is desired to communicate. Colour associations can convey the character of a product and might be used to convey, say, a 'garden fresh' image. Colours associated with parts of the home, for example pink with bedrooms, would be very suitable for packages likely to be used mainly in various parts of the home.

Fashion   Some colours have a high fashion connotation quite apart from current trends. Select colours for packages according to the image that it is desired to convey. The message may be one of fashion or high sophistication and the right colour can convey this irrespective of trends. Although black is not a good colour for packages, it might be used to convey high fashion in some cases; a strong trend colour (such as brown at the time of writing) would do very little for some products, such as cosmetics, because it would have the wrong associations.

Impulse   Some colours have high attraction value and are compulsive; they impel people to look at them. Select colours having a high attraction value for most packages and particularly when it is desired to create impulse sales.

Markets   The broad principle is that the mass market likes bright and simple colours; subtle and sophisticated colours are for higher grade markets, but there are also colours that are appropriate to specialised markets. Select hues for packaging according to the market at which the product is aimed.

Mood   Colour appeals to the emotions and different colours convey different moods. Appropriate colours may be used to set a mood or convey an image, for example of luxury. Select colours that convey the mood or emotion that the sales theme requires, such as freshness, softness, sunshine, excitement, good breeding, luxury.

Personality   Colour can be related to personality; specific colours appeal to people with well marked personal characteristics. Select only for novelty appeal as part of a well planned marketing scheme; it has been used for personalised stationery.

Preferences   There are general preferences for colours which vary with age and which should be distinguished from trends. Use preferred colours only when all other circumstances have been weighed.

Products   Colours are often associated with specific products either by tradition or long usage. Certain colours are particularly associated with products used in the home. Select colours for packaging which have the right association with the product and with that part of the home in which the product is used.

Recognition   Some colours have better recognition qualities than others, although not necessarily better visibility. In packaging, colours having good recognition qualities are significant in creating and enhancing a brand image and may have to have priority over all other considerations; they also help to attract attention.

Reflectance   Every colour reflects a different proportion of the light that falls upon it and colours having a high reflectance value are generally best for packaging; they also reflect heat and have other functional applications.

Regional   Certain colours are more appropriate to one part of the country than another, although this has limited significance in packaging applications.

Seasons   There is an association between colour and the seasons of the year, for example green and yellow for spring, brown for autumn. These may be used in packaging where useful and for seasonal promotions. Easter confectionery should reflect spring colours; summer colours might be used for sun-tan lotion; Christmas packages benefit from appropriate colours.

Sex Some colours appeal more to women than to men and most colours have sexual connotations. Use where a package is aimed primarily at one sex.

Shape Colours tend to suggest certain shapes in psychological terms and people seem to prefer certain colours in relation to a specific shape, for example blue is best suited to a cube. This may be useful in creating greater emotional appeal.

Size Colour can make objects look larger or smaller, heavier or lighter, nearer or further away. This is significant in many packaging applications, especially where promotion emphasises the size of the package or where the sales theme is value for money.

Smell There is an association between colour and smell and this is useful where perfume is part of the appeal of a product; the colour of the package will remind people of the aroma. A good, strong brown could be prescribed for coffee but would not be suitable for a delicately scented tea.

Stability Colour can be used to suggest stability, dignity and restraint and this may be useful in establishing a corporate or a product image.

Taste There is an association between colour and taste which varies with products. This is significant in packaging where flavour is an important selling point.

Tradition There are traditional colours and decorative ideas, such as Adam or Regency, which can be used where appropriate but have little place in packaging except in special cases, for example an 'old fashioned' product.

Visibility Certain colours have better visibility than others although not necessarily better recognition qualities. Colours having maximum visibility should generally be used in packaging applications. Visibility is particularly important if all promotion is at point of sale.

Warmth Warm colours are more appealing than cool colours and, apart from anything else, are useful in creating an inviting display.

3.4 Colour functions

Colour has a number of practical uses in all graphical applications, apart from its decorative value. Those that are relevant to packaging may be summarised as follows:

Coding Colour is frequently used for coding and identification and the colours need to be carefully selected to ensure maximum contrast between each one. In packaging, different colours are frequently used to distinguish flavours, scents, tastes and so on.

Protection from light   Colour can be used to protect the contents of a container from the harmful effects of light.

Readability   Some combinations of colours have better readability than others although they are not necessarily more appealing in emotional terms. The most readable combination is black on yellow but this is not necessarily the most suitable for a package; the impact on the viewer and recognition qualities are equally important. Black on white is highly readable but colour would have twice as much impact.

Temperature control   Bright colours tend to reflect heat away from an object and keep the interior of a package cooler. This may be a useful way of protecting the contents of a package.

3.5   Colour applications

There are a number of applications of colour which are significant in package design, including the following:

Brand image   The use of colour to establish a brand image or corporate identity is a complex subject but very important in package design.

Export   Colours that are suitable for British markets are not always suitable for overseas use; people of different races have different reactions to colour. Special consideration is necessary if a package is to be used overseas. See Part III.

Food   There is a definite art of colour in relation to food which is very significant in selecting colours for packaging. See Part IV.

Graphical display   This term refers to point of sale displays and the like and requires special consideration in package design where point of sale attraction is important.

Illustration   Where illustration is used in graphic design, special care is necessary.

Safety   Safety colours have limited applications in packaging except in labelling.

Tags and labels   Certain colours are particularly suitable for tags and identifying labels which require high visibility and the right association with the product.

Television   Colours react in different ways to the television camera and this is significant when selecting colours for packages which may appear on television.

4 Marketing decisions

4.1 The objectives

Promotional material, including packaging, is a marketing tool. Whether to use promotion, how to use it, and what media to employ are essentially part of the process of market planning and it is unnecessary to go into details here. The pros and cons of, say, press advertising over direct mail, or of cans over bottles, depend on what is being sold and how it is being sold.

At some point, however, it becomes necessary to plan the individual piece and a number of marketing decisions are required, particularly about the objectives to be achieved and the way that the target is to be reached. The primary objectives of all promotional material have been summarised in Section 1 of this part, but these primary objectives can be expanded further depending on the nature of the material. The ultimate object of most promotional material is to sell a product or a service, or perhaps to communicate an image or a message; the objective may be to trigger action, for example to persuade people to make a purchase or at least to make further enquiries, or it may simply be to provide information or create a reminder of an organisation or brand. With packaging, the primary objective is to sell a product, but there may be secondary objectives which will help to facilitate a sale, and the package may also help to promote a brand image.

Consider, in all cases

- the aim of the promotion
- what the piece is intended to accomplish
- the action sought – purchase, enquiry and so on
- which of the functions that the piece has to perform is the most important.

The following is a list of typical objectives, and the marketing plan should indicate which of them is to be given most weight in graphic design and colour choice:

- Attract attention. This is usually the most important function of all, particularly with packaging, where it is primarily a function of colour. With other forms of promotion, such as press advertising, graphic design may be more important.
- Persuade the customer to read. This requires visual appeal, legibility and the right combination of colour, and applies to packaging as well as to other printed material.
- Put over a message. Largely a matter of design and creative copy, but colour can help. Less important with packaging.
- Trigger sales. All elements must combine together to make an impact – the primary objective in the case of packaging.
- Remind. The combination of design and colour should be sufficiently distinctive to remember. Applies to packaging.
- Supply information. Legibility and broad visual appeal are important. Applies equally to packaging and to other material.
- Create desire. All elements must combine together to make an impact and invite further enquiries. An essential part of packaging.
- Create a mood such as excitement or nostalgia, thus leading to further action. Applies to packaging, and colour is particularly important.
- Arouse interest in fashion. Both colour and design must reflect current trends. May apply to packaging in appropriate cases.
- Create a desire for change. This is similar to creating desire and applies to packaging. It is a trigger which leads to further action.
- Create an image of a company, product, brand or service. The chief requirement is that colour and design should associate together in a way that will be remembered. Applies to packaging.

There may be other objectives in individual cases but the primary target to keep in sight is that promotional material, by definition, is intended to sell – and is more than just a pretty picture.

When the marketing decisions have been made, thought can be given to design and to briefing the designer. However, before doing so it will be useful to consider the image that it is desired to put over. All promotional material is a promise to the viewer, and in the case of a package the customer is asked to buy on the strength of a promise made by the message conveyed by the package. This is particularly important because people use appearance to make judgements about reality. The image to be put over may be a hard sell or a soft sell and it can take a number of different forms, for example

- flashy
- dignified
- sophisticated
- technical
- impulse appeal
- fashionable
- new
- up to date
- traditional.

Other ideas will readily present themselves.

4.2   Is colour necessary?

Before going any further it will be useful to decide whether colour is necessary at all. Although colour is an essential part of the marketing process, marketing is also concerned with economics and not necessarily with glamour; there is no point in using colour unless it pays dividends in the form of increased sales and helps to achieve the primary objectives of the whole operation.

The answer to the question will usually be decided by economic considerations, such as whether the increased sales will justify the additional cost of colour. This is a matter of commercial judgement or management decision, but it also depends on whether colour can be effective. This question is less likely to be asked about packaging than about other forms of promotional material, but it should be asked nevertheless.

There may be some cases where good black and white will have just as much impact as colour, and both can fail if the planning concept is wrong. The effectiveness of colour may well depend on the way that it is used and on the actual hues chosen; poor register, garish tones and outdated variations can kill the advantages of colour very quickly. There may even be cases where colour creates an unfavourable climate because the public feels that money is being wasted.

Consider

- the economic justification for colour. The cost is justified if it secures more attention or leads to greater sales.
- whether colour can create greater impact. Colour may be justified on decorative grounds or to add appeal.
- whether the message can be conveyed without colour.
- whether colour is essential because of the nature of the product. It may be necessary to convey the reality of what is available.

4.3   How is colour to be achieved?

This question follows on from the previous one and may have an effect on the economics of the situation. If good colour is to be used for most promotional material, the choice of medium will be significant and will be a factor in the marketing equation. The way that colour is achieved is equally important with packaging and will depend on the nature of the package, for example

- tinted board
- coloured inks
- process colour
- product illustration
- decorative illustration
- litho printed labels
- direct printing on cans.

The method of producing colour may well be a significant factor in deciding which colours to select; there is no point, for example, in trying to illustrate a product in colours which do not reproduce well on carton board. If an appeal is to be made by using subtle colours, the medium must offer a high quality of reproduction. An illustration of an expensive and sophisticated product would be wasted on a poor quality medium.
Consider

- what medium is to be used
- what limitations the medium imposes
- what limitations the process imposes
- whether colour has to be selected for good reproduction.

4.4   How is colour to be used?

Once it has been decided that colour has a part to play and it is known how colour is to be achieved, it then becomes necessary to decide on the right colour, or colours, and how they are to be used to best advantage. Choice of the right colour is primarily a marketing decision; the marketing plan should specify the colours to be used, and the reasons for using them, before the designer is asked to put them together in the most effective way. The reason for choice will not only affect the selection of the actual hues but may also dictate how the designer uses them in the finished design.

In reaching decisions it will be useful to consider what colour can do for the promotion under consideration. The answer will depend, of course, on the nature of the promotion and on the marketing objectives to be achieved. Some of the things that colour can do are listed in Section 3 of this part and it is recommended that the various attributes and characteristics of colour should be reviewed so that it can be decided how they can be used to best advantage.

The aim is to translate marketing and merchandising policy into colour terms in such a way that there is a clear understanding in the minds of all concerned, including the designer, about the objectives of the operation and how colour is to help to achieve them.

There are a number of special points to bear in mind:

- In any promotion it is desirable to select the special features that are to be emphasised and then to identify those attributes of colour which will underline the emphasis.
- When promoting specific products, select those features of the product that have most appeal to the customer and use the attributes of colour to underline those features.
- There must be co-ordination between various forms of promotional material having the same subject matter, for example between packaging and press advertising, or between press advertising and television. It is not always possible to consider each field in isolation.
- Special care is required where illustration is used, particularly illustration of a product. The colour of the product is the important factor and that will be selected because it has most appeal to the *customer* for *that product*, and not for graphical reasons. Subsidiary colours should be selected to support the product colour.
- Different considerations apply to promotion aimed at industrial markets; the nature of the customer is different and the appeal of colour may be practical rather than decorative. Trends in architectural decoration and office colour may be more important than consumer trends.

5   Which colours to use

5.1   The problem

When all the marketing decisions have been made and the aim and nature of the proposed promotional material have been clarified, it is still necessary to select suitable colours. Although colour can be used in many different ways, it is only the right colour or colours which will contribute to maximum sales or ensure the optimum benefit from the promotion. Colour can fail just as easily as black and white unless proper attention is given to selection. Colour for the sake of colour is not enough; there must be good reasons for selection, reasons which contribute something to the ultimate objective. For example, in addition to attracting attention, the colour may have to support a specific sales theme and, at the same time, be appropriate to the medium.

Much research into the subject of colour for promotion, and particularly into colour for packaging, is too academic and researchers tend to credit the spectrum with qualities it does not possess. The most common fault is to place too much emphasis on more colour and not enough emphasis on the right colour. There is also a tendency to be too clever and to try for effects which are beyond the comprehension of the average person.

The problem of those responsible for colour selection is how to identify the right colours; selection of colour must not be too mechanical, nor should it be made on purely artistic grounds. The colours selected not only have to create visual interest but must also have lasting appeal, and this depends on many factors which have to be assessed in each case. It is necessary to reconcile optical and physiological facts with emotional pleasure and to strike a balance between visibility, impact and emotional appeal.

There is very little point in using colour if it is not employed in the most effective way. Although this is partly a matter of good graphic design, choice of suitable colours cannot be left entirely to the designer; the latter should be willing to accept that certain colours have advantages from a marketing point of view and that hues should be selected for marketing reasons. However much care is devoted to the planning, design and preparation of promotional material, the effort will be wasted if the public does not like the colour. The effort will be equally wasteful unless colour strengthens the impact of the promotion and performs a useful function.

Whatever the nature of the promotion it has to appeal to people, and their likes and dislikes will determine whether the piece is effective or not. The colours have to have sufficient impact to attract attention in the market situation for which they are intended, and should have the right association with the product being sold or with the aim of the promotion. Colour should be used with purpose; never use colour simply for the sake of colour. In addition, the appeal is affected by trends of consumer preference for colour, and the significance of colour cannot be properly appreciated without a clear understanding of the influence of trends.

It follows that the facts of each situation have to be analysed and basic data brought to bear on the outcome. Use basic facts first and then good taste and common sense.

Colour selection is a complex process and if carried out properly it involves a good deal of work. It is essentially part of the marketing process; it involves analysis of the whole marketing picture and isolation of those functions that colour can, and has to, perform. The purpose is to identify the right colours for a particular situation, and this depends on many factors including the product, the nature of the market, the marketing plan and the media used. It also involves identifying those attributes and characteristics of colour which will be helpful in the situation under consideration, and selecting hues which have the appropriate qualities.

The whole may require a combination of marketing research and colour research techniques and a study of the physiological, psychological and optical influences of colour. The end result should be a colour or a combination of colours that will attract maximum public acceptance and will ensure that sales are not lost because selection has been based on guesswork or the prejudices of the individual.

The first step in the process is to define the marketing aim and to pinpoint the nature of the market concerned, and the requirements of that market; on this depends the quality of the hues concerned. However, selection also depends on a number of other complex factors such as the type of customer, the nature of the product and the nature of the sales theme. Colour may be selected specifically to support a sales theme.

A second step is to identify those responses that it is desired to trigger off. The basic hue may be selected for physiological reasons, such as to attract attention; for psychological reasons, such as to convey a mood or an impression; or for functional reasons, such as to provide identification or readability. Even when a suitable primary colour has been pinpointed, each primary hue has thousands of shades, tints and tones. It is necessary to take heed of trends in consumer preference to hit on the variation that will be current at the time when the promotion is seen by the public. The colours that people like change over the years and preferences vary from time to time, both overall and in relation to specific products.

The resulting selection may be subject to limitations imposed by the material used and by the method of reproduction. Further limitations and modifications are necessary when the piece includes a product illustration; the colour of the product will then be the dominant factor.

It will be appreciated from these brief remarks that the colour selection process can be complicated, but it is possible to adopt a systematic approach which simplifies the sequence of operations. This is discussed in the next section. The initial problem is to decide what information is necessary and how much research will be required to provide that information.

Consider what is required to

- identify the market
- isolate the attributes and characteristics of colour appropriate to the marketing aim
- establish what is necessary to support a specific sales theme
- identify the responses to be triggered off
- identify trends
- identify other factors affecting selection, including limitations.

5.2   Solving the problem

As indicated in the previous section, the colour selection process is complex. It can be simplified by adopting a systematic approach to the whole subject, making maximum use of accumulated knowledge about colour and applying it to situations which have been clearly defined by means of research. In adopting a systematic approach, the colour selector should think about colour and use intelligence and common sense rather than guesswork, hunch or bright ideas. The fatal mistake is to introduce personal preferences; creative people, in particular, are apt to favour clever effects instead of thinking about the purpose behind the use of colour.

The right colour for any promotional material is the right colour for the job that the material is supposed to do, and this has to be established at the outset, it is a marketing decision. No designer can give of his best without a clear idea of what is required, and the same applies to colour selection. The whole should be planned to achieve a specific marketing aim, and the first step is to clarify the nature of the promotion and then to approach colour selection in a systematic way.

It is recommended that *a colour specification* should be prepared for each situation, setting out all the various factors involved, the market conditions concerned and the attributes and characteristics of colour that are appropriate. This is not a simple task, but there are many factors which are or may be involved, and the effort is well worth while because it obviates the risk of any important factor being overlooked and simplifies the whole approach. For example, it may be taken for granted that the form of promotion is press advertising, packaging or whatever, but forgotten that the same design may be used for more than one medium or, at least, that there must be co-ordination.

Marketing is the key to the whole operation, and the specification must clearly define all the marketing and other considerations which will apply to the situation under review; they will, of course, vary with the object of the material.

The basic steps in the preparation of a specification are as follows:

Decide what colour has to do   This requires a complete analysis of the nature of the promotional material and the functions that it is designed to perform, including

- the marketing aim
- the nature of the customer at whom it is aimed
- the message that it is desired to communicate
- the conditions under which the material will be seen.

Decide what colour can do to help achieve the objectives and which attributes and characteristics of colour are appropriate to these objectives.

Establish the marketing picture   This is a most important step and provides the answer to some basic questions which affect the whole planning process. Some of these questions depend on the characteristics of the promotion and some depend on management decisions. The aim is to define the nature of the market quite clearly so that all concerned know where they are going and how colour is to help in the process. The end result of this stage should be the development of a colour strategy.

Consider the nature of the problem   This stage includes consideration of the media to be used and the method of reproduction because these affect all that follows; it is not any use selecting a colour which cannot be reproduced. It also includes deciding what steps have to be taken to provide information about the characteristics of the market and the factors affecting colour choice. It would be advisable to establish machinery to keep track of trends.

Identify the market   Define the nature of the market at which the promotion is aimed, and the type of customer that it is desired to attract. The make-up of the market may have influence on colour, as may the activities of the competition. The characteristics of the product may be important and so may selling conditions and the purchasing habits of the customer.

The sales theme   Consider the sales theme, or sales message, and how colour can be used to support it. Seasonal factors may be significant and appropriate colours are prescribed.

Other considerations   Consider the various functions and applications of colour and how they may be used to best advantage. Readability is particularly important for most promotional material.

When all the required information has been collected and the specification is complete, it is possible to select actual hues whose attributes and characteristics will best fulfil the conditions identified in the specification. The first crude selection will have to be modified to reflect current trends of preference for basic colours and for variations of basic hue. Whatever the reason for selection of hue, use the current popular variation of that hue.

When the right hues have been found they have to be used to maximum effect and this requires attention to colour modifiers, colour combinations, the rules of harmony and the most effective arrangement of colour.

6 Identification of the market

6.1 The product

Selecting colour for any promotional material is a combination of marketing planning; of consideration of the attributes of each hue; of consideration of colour trends; and of a judicious weighing up of the advantages of colour, plenty of colour, and no colour at all.

Before a specification can be drawn up it is necessary to identify the nature of the market at which the piece of promotional material is aimed. This will almost certainly require marketing decisions and may require extensive research into the various features of the market; until these have been clearly established it is difficult to select colours, because the attributes and characteristics of individual hues have to be matched to the requirements of the market.

The first point to consider is the influence of the product itself in those cases where the promotional material is concerned with a single product; the nature of the product will obviously have a considerable influence on the selection of colours used in promotion and the choice will depend very largely on whether the product is illustrated. If it is, the product will usually appear in its natural colours or in the colours in which it is sold. These colours will, in turn, govern the hues used for background and for other features.

Selection of colours for products is no part of this report. Where the product is produced in a variety of colours, management will decide which variation they will use in promotion, using criteria of choice which are different from those applicable to graphical material; normal practice is to feature the best selling colour. Where colour is the main selling feature of the product, as with carpets, always illustrate the product in colours that people will want to buy. It may be necessary to illustrate a range of colours.

When the product is not illustrated, or when the product colour is not a vital selling point, the background may be especially important and the colours used should associate well with the product; this is particularly important in the promotion of foodstuffs.

Consider the following points:

- Are there any colours associated with the product? Many of these are traditional and have grown up over the years.
- Are there any colours which will enhance the product in the eyes of customers? For example, pink conveys sweetness.
- Are there any specific likes and dislikes in relation to the product? There may be preconceived ideas about the right colour.
- Is there any colour which will enhance the principal attributes of the product? For example, dark green is associated with pine odour.
- How do customers buy the product? If it is bought on impulse, the colour of the packaging or of the point of sale material is important.
- When do most sales take place? Seasonal sales require an appropriate background colour.

6.2   Services

When promotional material is concerned with services or with communicating an image, there is no product colour involved and colours will usually be selected because of their association with the service or because they create an image or a mood. The nature of the market and the type of customer it is desired to attract are still important.

With travel promotion, for example, colour will be used to convey the attraction of a holiday, sunshine, the sea, the outdoors and so on; sunny, bright colours are indicated. Promotion to business men, such as financial services, requires a more restrained image and colour should be influential and quiet in nature.

It is difficult to formulate precise guidelines, and intelligent thought is required. Consider

- what image it is desired to convey
- what mood is indicated
- any colours associated with the service
- whether readability is particularly important.

6.3 The market

It is necessary to pinpoint the market at which the promotion is aimed; it may be aimed at all markets in which a product is sold, or it may be aimed at a specific market segment, depending on marketing plans. Promotion might be aimed at a specific market sector as part of a policy of trading up.

Promotion of a service may also, of course, be aimed at a specific market and, while the same broad rules apply, careful consideration must be given to the target. Brash colours suitable for the mass market would not have the same appeal to business men.

In broad terms, the mass market likes bright and simple colours but higher grade or sophisticated markets prefer more subtle tones. Experience has shown that the top 10 per cent of the market likes to be different to anybody else and will tend to be attracted by 'different' promotion. It would be useful to use more sophisticated variations when trading up; 'supermarket' colours would not be appropriate in such circumstances. If the promotion covers a product or a service sold on snob appeal, it would need quite different treatment compared with a mass market product and would benefit from more sophisticated colour and a higher degree of fashion. A high fashion market merits particularly careful treatment; certain colours convey an impression of fashion and should be used where they are appropriate.

Consider the following:

- In consumer markets, generally use trend colours.
- In business markets, select colours for practical reasons.
- In technical markets, select colours for functional reasons.
- In upper class markets, more sophisticated variations are recommended.
- In mass markets, bright, simple colours are preferred.
- In replacement markets, use restrained colours.
- In fashion markets, use up to date fashion colours.
- In export markets, consider each market individually.
- In local markets, consider prejudices which may have to be discovered by market research.
- In all markets, consider any prejudices or fixed ideas that there may be.

Special care is necessary with some trades. For example:

- In textiles, pattern as well as colour is important.
- In clothing, up to date fashion colours are recommended.
- In furnishing, background to illustration is significant.
- In carpets, colour is essential.

6.4   The customer

The nature of the customer at whom any promotion is aimed will have a significant influence on selection of colour. In many cases the average customer is a clearly defined section of the total market and in other cases promotion will be aimed at a specific type of customer. In all cases, colours should be selected to have maximum appeal to the major market.

The make-up of the customers in terms of age and sex is important. Promotion aimed primarily at the youth market should be bolder and brighter than material aimed at older and more traditional customers. Where products are concerned, there is no point in aiming at wild youth if the customers are middle aged married women; this would be selling in a market which did not exist.

Certain colours have greater appeal to young people than they do to older people; therefore it is important to discover whether the greater proportion of customers are in young age groups or whether there is a more universal target. A product or a service may be specifically promoted to young people as a matter of marketing policy. Colour should always be selected to have maximum appeal to the age group concerned.

The sex of customers is also significant. Women, as a general rule, react favourably to more gentle colours than men, and certain colours (such as pink) are distinctly feminine in appeal. This is quite different to the use of sex as a basis for advertising; a minority of women are unhappy at the way that women are portrayed in advertisements and sex may well turn off more customers than it turns on.

The overall preferences of ordinary people apply when colour is used for the sake of colour and without other motivation. Red, blue and green are the most popular colours overall among adults, although children prefer yellow. However, in all cases, current trends in consumer preferences must be considered; if people like colours to live with, they will also like them in promotional material. It is particularly important to use up to date variations whatever basic hue is used.

Consider the following:

- Who is the customer?
- Whom is the promotion intended to reach?
- What colours are appropriate to the major customer group?
- Which sex is the major customer?
- Which age group is most significant?

6.5   Purchasing habits

The purchasing habits of the average consumer are not a significant factor in the selection of colours for most promotional material. However, they are very significant in the case of packaging, and display material, when sales may be made on impulse; in such cases it is essential that the colours used should have maximum attraction value. They may also be significant in the case of a considered purchase where purchasers will be favourably impressed by fashion colours or by a specially attractive colour scheme.

Consider

- the purchasing habits of the customer for the product
- whether colour is significant
- impulse colours where purchases on impulse are indicated
- special colour schemes in special cases.

6.6   Selling conditions

Selling conditions have little significance as far as most promotional material is concerned, but they are a vital factor in the selection of colours for packaging and point of sale material. In both the latter cases, the conditions under which a piece will be seen are very important and should be reviewed in detail. A point of sale display, for example, should contrast with its surroundings and the possible effect of after-image should be considered.

In packaging applications, supermarket selling conditions may be very different from those in other types of outlet, and quite different treatment may be justified for packages intended primarily for supermarkets. Display conditions often vary with different systems of distribution and this may dictate some modification of colour choice. Coding and identification may be more important under some selling conditions and it may be necessary to distinguish between flavours or textures.

The lighting conditions under which a piece is seen may also be very significant, and this is discussed in Section 9.5 of this part.

Consider

- selling conditions generally
- the surroundings in which a particular piece will normally appear
- the significance of supermarket selling conditions
- the significance of distribution arrangements
- the need for coding
- lighting.

6.7   Usage

The usage of a product normally has little influence on colour selection for promotional material except in packaging applications, where the use to which the package is put after sale may have an influence on colour choice. For example, a container of salt which is designed to be used in the kitchen, or on the dining table, would be appropriate in colours that are acceptable in the kitchen of the average home.

Consider

- the usage of the product
- possible influence on colour selection
- after-use of packaging.

6.8   Competition

It is very useful to decide who, or what, is the competition and to study their use of colour. What colours do they use? What theme do they use? Has their use of colour any marked advantages? It may be necessary to change colours to avoid being overshadowed by the competition or to avoid plagiarism.

Consider:

- Who is the competition?
- What is the competition?
- What colours do they use?
- What advantages do they have?
- Do they dictate a modification of colour choice?

6.9   Sales theme

Colour has meanings for people, and it can be used in promotional material, including packaging, to express the message that it is desired to communicate. Such expressions as 'garden fresh' or 'mountain cool' can be emphasised by appropriate colours, and in some cases colour can be used to suggest moods or themes without actual use of words. It may suggest the character of a product, or the promise of a service. It may be part of the image of a product.

Even when colours are selected for other reasons, the colours ought to be checked for their possible effect on the image of the product. For example, black might be included in a design because of its association with sophistication, but in the wrong context it might convey a negative image because of its association with mourning. The marketing plan should take cognisance of the time of year that is appropriate to the product and use appropriate seasonal colours for the design on its package.

Consider

- the message that it is desired to communicate
- the selling features to be emphasised
- the image that it is desired to suggest
- whether seasonal factors are significant
- how colour might be used to support the sales theme.

7 Other considerations

7.1 Export

If the promotional material is to be used overseas, special care is necessary with the selection of colours. Some colours are unsuitable for overseas markets because of local likes and dislikes. People living in sunny climates generally prefer lighter colours than those who live in temperate climates; promotion and packaging has to compete with a brighter environment. There are also prejudices such as that against green in Moslem countries. Colour associations and trends may differ from British markets.

Consider

- each country in which the material will be used
- colours which are appropriate to each country
- the need for different colours for different markets.

7.2 Brand image

Colour can be used to create and identify a brand image or corporate identity and, where this has been done, the appropriate colours should always be used in promotion. It is important to maintain standards.

It is not practicable to say that one colour is better than another for this purpose, and colours should be selected with great care and have the right association with the product or service.

Consider

- the whole question of brand and corporate image.

7.3   Merchandising

Some colours, by reason of their characteristics or optical properties, are more suitable for background, while others are recommended for attracting attention. This is particularly important in display applications, and there are many practical ways in which colour may be used to enhance display.

Any promotional material which is to appear at point of sale, including packaging and point of sale display, should be reviewed in the light of the part that it might play in merchandising. It is difficult to formulate any hard and fast rules.

Consider

- the implications of merchandising.

7.4   Food

Promotion of food requires special consideration because some colours associate well with food while others do not. Food packaging requires special care and so does food illustration.

Consider

- the association between colour and food, and especially colour and food packaging.

7.5   Tags and labels

Certain colours are particularly suitable for tags and identifying labels which need the highest possible visibility and the right association with the product. This is a specialised form of packaging and deserves special study.

Special care is necessary with warning labels, which are often printed in red; red becomes virtually invisible in some lighting conditions.

Consider

- whether tags or labels are necessary
- the most suitable colours for the message to be displayed.

7.6   Television

Colours react in different ways to the television camera and special care is necessary in selecting colours for promotion which will appear on television. It is, for example, important that a package which is advertised on television should appear exactly the same on the screen as it does when seen on the shelf. Any piece which appears on the television screen should be easily recognisable when seen away from the screen.

Consider

- the implications of television exposure.

7.7 Coding

Identification and coding is a function of colour and will only be significant in promotional applications when colour is used to distinguish one piece from another. For example, packages of a product may be colour coded to indicate different flavours; leaflets may be colour coded to indicate differences in subject matter; catalogues may be colour coded to indicate different product headings.

It is usually desirable to use primary colours for coding purposes and to ensure that there is maximum contrast between one colour and another. It is difficult to say that one colour is better than another for this purpose, but strong hues are generally better than pale ones.

Colours used should generally associate with the subject matter. A typical example is the association of green with peppermint.

Consider

- whether coding is necessary
- maximum contrast between colours
- association of colour with product (or service) where practicable.

7.8 Protection from light

Another function of colour is to provide protection from light, and this chiefly applies to packaging. A typical example is the use of brown bottles to protect beer from the effects of sunlight, but coloured films are also frequently used to protect the contents of a package.

Where protection from light is desirable, it would be advisable to study the subject in some detail because the most effective colours from a protection point of view are not always appealing from a marketing point of view.

Consider

- whether protection from light is desirable
- which colour will best achieve the purpose.

7.9   Temperature control

This is a minor function of colour in graphical applications but it may be of some significance in packaging, particularly packaging for export.

When the outside of a package is finished in light colours, these reflect heat and light and tend to keep the contents at a more equable temperature. This may be useful when a package is displayed in strong lighting conditions and when the contents might easily spoil if subjected to heat. The use of aluminium or other shiny foil serves the same purpose.

Consider

- whether temperature control is useful.

7.10   Safety

There is a recognised code of safety colours to denote fire appliances, first-aid points and so on, but these have little application in promotional material. However, the use of colour for safety purposes may be significant in packaging and label applications. Warnings and the like, whether required by law or not, should certainly appear in bright, visible colours and there are accepted colours and symbol combinations to denote specific hazards. These should be followed where appropriate.

Consider

- whether there is a safety element
- how colour can improve safety
- whether there is a recognised safety colour.

7.11   Readability

One of the most important functions of colour in all promotional applications is to ensure readability and legibility and to promote the desire to read. Although legibility is mostly a matter of type size and layout, colour also has an important part to play by ensuring good contrast between legend and background and, at the same time, providing emotional appeal.

The most legible combination of colours is black on white because this provides maximum contrast, but it lacks any emotional appeal. White can provide a good background for almost any colour but it is not the only possibility, and other colours can be used provided that they contrast with the type colour. As a general rule the legend should stand out and the background fade away because this makes the legend easier to read. The reverse is sometimes used in the interests of novelty, but 'clever' effects cannot be read in many cases.

Certain combinations of colours are more readable than others because of the reflectance ratio between the colours. Black on yellow is particularly good from a legibility point of view but it lacks emotional appeal and is not very attractive in promotional applications. A combination that has good legibility and emotional appeal at one and the same time is far better in most cases, although extreme legibility may be the first consideration and the likes of the public may have to take second place.

Consider

- the importance of legibility
- how legibility is to be achieved
- the most appealing combination.

8 About trends

8.1 General considerations

An ideal situation exists where colour does its job efficiently in a visual sense and pleasantly in an emotional sense at one and the same time, and to achieve emotional pleasure colour should reflect current trends. That trends of design and colour are a vital part of the ideal package has been explained in Part I. They are an equally vital part of any promotional material, but it may not be clear why colour trends are so important without an explanation of their nature.

There are a number of different types of colour trend, for example in

- consumer preference for colour in the home
- colour in men's and women's clothing
- colour usage in the office
- colour in architecture.

There is a relationship between all these trends. However, it is the first that is important in promotional and packaging applications because it reflects the wishes and desires of ordinary people and influences everything that they do. Trends of colour in clothing are only important in an application directly related to clothing.

In many cases promotional material and packaging should reflect current trends in consumer preference for colour and should certainly reflect current popular variations of basic hues. If people like a colour sufficiently well to purchase it for use in their own homes, they will also like it when they see it used in promotion, or on a package on the shelf. It is the same people who are involved at every stage; they see the advertisement, they are motivated by the packaging and they buy the product. While every individual has his or her own personal preferences for colour, it would be quite impossible to give effect to these personal preferences, nor is it necessary. The great majority of people tend to like the same colours at the same time – hence trends.

In addition to preferences for specific hues, there are also preferences for types of hue, for light colours, for bright colours, for pastels and for muted hues. These preferences change periodically on lines similar to, but not parallel with, preferences for hue. The selection process embraces brightness of hue, as well as shade, tint and tone.

Trends are particularly important where product illustration is involved because people expect to see the same contemporary colours in promotional material that they see in the products that they buy at point of sale.

Consider

- whether trends are important to the product being promoted
- how trends can be reflected in the material
- what the current trend colours are
- whether trends favour bright, pastel or muted variations.

8.2 Variations

The initial stage in the colour selection process is to identify those hues that have the right attributes and characteristics for the assignment in hand. This may result in a number of possible alternatives, and it would be desirable to consider these alternatives in relation to current trends of consumer preference for colour in the home. It is always sound practice to use current trend colours if at all possible, particularly where a product has a fashion note, or where it is used in the home. If people like a colour in their homes, they will also like it on a package.

For example, a household appliance which has no particular colour associations would benefit from using current trend colours for its packaging; if the appliance is used in the kitchen, or is associated with the kitchen, use kitchen trend colours.

Whether or not it is practicable to use a trend colour, it is always practicable to use a variation which is popular by current standards. At any given time a bright variation may be preferred to a muted one or a strong colour may be preferred to a pastel, as explained in the previous section. A variation chosen at random from a standard colour range may be obsolete and do more harm than good. A variation which is popular or fashionable will have more emotional appeal than one that is out of date.

For example, red might be chosen for a package because it is a colour of high impulse value, has good recognition qualities and has a favourable association with the product. Red, as such, may not be a strong trend colour at the time but there will almost certainly be a preference for one variation of red as compared with others. Yellow type reds might be preferred to blue type reds and therefore best for packaging applications. However, preferences change and at some other time blue type reds might be a better bet. Red is a particularly good example because there are quite strong preferences for different variations of red at different times.

Although people will always be attracted by red because of its physiological attributes, they will also be psychologically attracted by the variation that they prefer for use in their own homes, and which they see around them in the things that they buy. An obsolete variation may suggest that a package is out of date. Similar remarks apply, of course, to all hues.

Most packaging and other promotional material is aimed at the consumer and therefore human emotions and preferences are vital if the packaging is to be effective. There *are* vogues for colours and these should be reflected in anything aimed at the consumer. The same consumers expect a change from time to time. It may not be practicable to change the basic hue if it has been selected for practical reasons, but it is usually possible to change the variation and bring it up to date.

The direction of trends needs to be kept under review by anyone concerned with the selection of colour for packaging, and if there is a strong fashion element involved it may be necessary to set up elaborate machinery to keep track of trend movements.

Consider

- which variation of basic hues to use
- whether change is desirable.

9 Usage of colour

9.1 Colour modifiers

When colour specifications have been completed and when suitable hues have been selected for a piece of promotional material, the way that the hues are put together is a matter for the designer. However, there are a number of phenomena which may affect the end result and which can be used to improve that result. These phenomena are known to most good designers but they are listed below for convenience. They derive from the fact that in all graphical applications light does not travel directly to the eye but is reflected light; the piece is perceived as having shape, size, texture and colour because of the way that light is reflected from what is seen. This reflected light can be modified in a number of different ways.

Most rules of colour harmony are based on the 'colour circle' in which the primary colours of the spectrum are arranged around a circle in the order red, violet, blue, green, yellow and orange. Thus red is the opposite of green and is described as the complement of green. Colours which lie next to each other on the circle, for example red and violet, are called adjacents, and intermediate colours are those which lie between two primary colours; for example blue green lies between blue and green. These terms have been used in the paragraphs defining colour modifiers which follow.

Adaptation The interpretation of colour may be affected by the state of adaptation of the eye at the time that an object is seen, that is the ability of the eye to adapt itself to different levels of light. This is primarily of significance in connection with point of sale material where visibility depends on the conditions under which the material is seen. A simple example is a tag which has to be seen in a dimly lit basement; in such cases avoid red.

- Consider the conditions under which the piece may be seen.

After-image Colour perception may be affected by what the eye has seen immediately beforehand. The consequence of this may be that an otherwise effective design is spoilt because the after-image of the colour on one part of the design alters the visual characteristics of the other part. However, after-image can also be used in a constructive way when perfect opposites are utilised, one against the other; the after-images of the two will enhance each other, provided always that the areas of colour are reasonably large. On the other hand, if two colours are confused or visually blended, the reverse may apply and the two colours will cancel each other out. When two colours are not exact complements, both may appear to be slightly modified in aspect.

- Consider the effect of one colour on the other when two or more colours are used in conjunction.

Colour blindness Colours may be seen differently if the viewer does not have normal colour vision. This is only of significance where colour is used for identification purposes or, for example, as a safety warning.

- Consider whether colour blindness is significant.

Colour constancy This concerns the ability of the eye to see colours as normal under widely different conditions of light and to 'remember' colour. Certain colours closely related to the experience of the observer become 'fixed' in the memory. Typical examples are the colour of the human complexion, of foodstuffs, of natural objects such as flowers; they are familiar and the brain 'remembers' them as seen under natural light. A designer using colour to express a brand image or a corporate image should aim at obtaining the same memory and, because of this, the colours used for the image should be controlled between very close limits. If distortions creep in they can be disturbing and reduce the value of the image.

- Consider the use of colour constancy in connection with brand or corporate image.

Contrast The appearance of colours may be affected by other colours within the field of view. Contrasts may be used in graphic design to enhance the appearance of a colour by putting it beside, or on, its exact opposite. Black print on white paper provides maximum contrast and maximum visibility; contrasting two opposites such as red and green will create attention. The principal features in any promotional design should have maximum contrast with their surround, and print should always have maximum contrast with its background to ensure maximum readability. In suitable cases the effect of after-image will enhance the contrast and create excitement; this is just what is required to attract attention. Contrast is particularly effective in large areas. There can also be contrast between light and dark; dark colours tend to accentuate the brightness of light areas and a light background will add depth to dark colours.

- Consider how contrast can be used to best advantage.

Illusion   The appearance of a colour may be altered by changing reality to such an extent that people 'see' things that are not there. In the graphic design context, this means modifying the appearance of a colour by the way that other colours are placed in relation to it. For example, a line of white type placed against a pure colour will seem to advance, thus adding an illusion of depth which increases impact. Illusion calls for the creative skill of the good designer.

- Consider how illusion can be used with advantage.

Juxtaposition   Colour appearance may be affected by light reflected from other surfaces within the field of view or in close relation to it. In all graphical applications, each colour or design loses something to each of the other colours in the field of vision and it is modified accordingly. The appearance of each colour inclines towards its adjacent (on the colour circle) which is furthest away from the colour with which it is in combination. Thus, if orange and green are seen together, the appearance of the orange would incline to red, and the appearance of the green would incline to blue. This effect is particularly noticeable in colour printing. In display and packaging applications, the colour of the surround or background should be carefully chosen so that there is no undue emphasis on any part of the spectrum. The background should be carefully related to the feature colour.

- Consider the possible effects of juxtaposition.

Metamerism   The appearance of colour is affected by the chemical composition of the surface on which it appears. The significance of this in graphic design is that a colour specified in, say, artist's colours or printing ink may look quite different on different substrates such as carton board and plastics film. Even different 'makes' of carton board may have an effect on reproduction, and so will different types and weights of paper.

- Consider the various substrates on which a design may be printed.

Perception   An image or a colour may be affected by the interpretation that the observer places on it; it may be a perception which exists entirely in the mind. In graphic design this usually means that it may not be possible to obtain a required effect within the limitations of the reproduction process. For example, it is possible for the artist to visualise a lustrous red but it may be much more difficult to reproduce this by commercial processes.

- Consider the significance of perception.

9.2 Combinations

Colour combinations are particularly important in graphical applications because of the need to achieve good readability, as indicated in Section 7.11 of this part. However, some colour combinations have more psychological appeal than others, and those with maximum visibility and legibility may not be pleasing to the viewer when used in promotional material.

Formal rules of harmony do not always reflect the feelings and emotions of ordinary people and it may be better to sacrifice some legibility in the interests of better appeal. Attention is drawn to the following section about harmony; lists of recommended combinations are available (see colour index, Part V).

Consider

- whether legibility or appeal is the more important
- what colour will best combine with the dominant hue
- recommended combinations.

9.3 Harmony

Most designers will be aware of the natural laws of colour harmony. However, they may not realise that these do not take into account the feelings and emotions of the average individual, and it is the latter which are important in graphic design for promotional purposes. It follows that formal rules should not be interpreted too literally or too arbitrarily.

The most popular colours among people at large are blue, red and green, and any combination in which they dominate will satisfy the majority of people. Opposites like red and blue and green are better liked than opposites like yellow and violet.

In most promotional applications it is a good general rule to choose a colour combination in which the current trend colour is dominant, but the dominant colour may be chosen for other sound reasons.

It is best to follow broad principles rather than strict rules, and the following are useful:

- All pure colours will usually harmonise with black and white.
- Tints of all kinds will harmonise with white.
- Shades of all kinds will harmonise with black.
- Tones of all kinds will harmonise with grey.

The majority of people find pleasure in colour combinations based on adjacent hues or opposite hues; intermediate hues are less popular.

Pure hues should always be vivid and intense and not indefinite. Pure red with a touch of white looks faded, and with a touch of black looks dingy; neither will be liked. If enough white is added to turn red into pink, or enough black to turn red into maroon, the results will be better liked because the pure hue has been turned into a precise tint or shade. Tints should be light and clean, shades should be rich and deep, tones should be mellow and greyish, and it should not be possible to confuse the form with any other form.

Strong, pure hues look best when they are confined to small areas and when they are seen surrounded by larger areas in shades, tints and tones. However, this rule may not apply in promotional applications, especially when red is used as an impulse colour. The nature of red is such that it ought to command a design and be emphasised through the use of small touches of blue and green.

Consider

- the rules of colour harmony
- the need for colours that appeal
- broad principles.

9.4   Identifying colour

Defining colour is a problem that presents considerable difficulty and there is no standard that can be used in every case. There are various charts and systems that may be used and there are systems developed for specific uses of colour. In printing, colour is often defined in terms of the standard colours of a printing ink manufacturer, but this may not always be practicable.

A term like 'light red' is never sufficient. The colour must be defined much more exactly and it is often best to work on actual samples which can be matched exactly by the printer.

It is not proposed to develop this subject further here but simply to draw attention to the fact that it exists and requires care.

Consider

- how colour is to be specified.

9.5   Lighting

Colour is, of course, affected by the nature or spectral quality of the light source. The chief significance of this as far as graphic design is concerned is that the piece may be seen under different conditions of illumination and each of the common sources can produce a different reaction from the same hue. For example:

- Incandescent light can make colours look more yellow.
- The effect of fluorescent light depends on the type of tube.
- Mercury light makes colour look more purplish.

Fluorescent inks are now commonly used in printing display material and packages and the effect of lighting on these can be even more pronounced; they only achieve their full effect in daylight and look particularly drab under sodium lighting.

There is no real answer to this particular problem because most material is likely to be seen under many types of light, but it would be desirable to select the colours so that they are most effective at the point where they have to make maximum impact. Thus, in the case of most packages, they have to make maximum impact on the supermarket shelf and it is recommended that the colours used should be those that show up best under typical supermarket lighting. If the piece is designed specifically for use under one type of light, adjust the colours to that light to secure maximum effect.

In case of doubt test out the combinations of colours under various types of lighting.

Consider

- whether the piece will be seen under all types of lighting
- the lighting under which it has to make maximum impact
- whether the piece is designed for a specific type of lighting
- whether the colours will realise their full value under normal lighting conditions
- the need to test colour appearance.

9.6   Light fastness

This is a property of pigments rather than of colour, but it is closely allied to lighting. Light fast qualities are very important where a piece of material is exposed to light, and particularly daylight, for long periods; point of sale displays and posters would be typical examples of material where colours should have good light fast qualities.

Some pigments and printing inks are particularly prone to fading and it would be better not to use them for material which will be exposed;

others have better qualities, and some are hardly subject to fading at all. It is usual for paint and printing ink manufacturers to quote the degree of light fastness of the colours that they offer and, before deciding on a particular hue, it would be advisable to check that it is light fast to a degree suitable for the job that it has to do.

Depending on the pigment or ink to be used, it may be necessary to modify a selected hue to ensure adequate light fast qualities.

Consider

- whether light fast qualities are required
- the qualities of the pigment to be used
- whether modification is desirable.

9.7   General rules

The following are some broad rules of colour usage which will be found useful in graphical applications:

- The eye prefers simplicity in colour just as it does in shape.
- Hard colours always persist over soft ones.
- Bright colours are preferred to deep ones because they afford greater stimulation to the retina of the eye.
- Colour will always dominate neutrality.
- Hard colours have better visibility than soft ones because visibility is dependent on brightness differences between two colours, irrespective of hue.
- Bright colours are more visible against a neutral background than the reverse.
- Bright, hard colours make an image look nearer to the eye but it is best to avoid one dominant colour because it is monotonous; contrasting colours will raise eye interest. The hard colours focus at a point behind the retina of the eye and thus bring colour forward; soft colours have the reverse effect.
- The brighter the colour, the larger the image will appear to be, because when brightness strikes the retina it tends to spread out like water on blotting paper and thus produce a larger image.
- Pastels look larger than shades because they are brighter.
- Pure colours are preferred to greyed ones because the eye and the brain respond more readily to pure, simple colours; they stimulate the retina to a greater extent.
- Greyed and other modified colours are not as well liked as pure colours; they tend to be depressing.
- If basic colours are modified towards white or black it is best to use precise steps rather than slight variations; this produces better contrast.

- Pale colours look light in weight; dark colours look heavy.
- Colours show up better when represented by a figure or shape than when used as ground. Soft or neutral colours are best for background.
- Brightness will accentuate darkness, and the reverse, because there is good contrast.
- Use deep colours in small doses; too much deep colour may be depressing.
- Use a colour of high visibility on a neutral background.
- Concentrate on the intuitive elements of colour. Sophistication may be appropriate in interior decoration but is of secondary importance in most promotional applications.
- The right proportions of colour create a balanced design.
- When a symbol is used, the symbol and its supporting elements should be in strong colours on a neutral or soft background.
- The eye tends to focus on an area slightly to left of centre and this is the dominant position, particularly for a symbol.
- Because the eye moves from left to right, the dominating figures should be on the left and less important features on the right.

Any graphic design used for promotion purposes involves the use of combinations of colours, contrasting colours, variations of basic hues, trends and other factors which may modify colours selected on marketing, psychological and physiological grounds. It is necessary to strike a balance between visibility, impact and emotional appeal and to reconcile hard facts with emotional pleasure.

Consider

- whether hard or soft colours are required
- shade, tint and tone, not only the primary hue
- whether simple colours can be used; they are almost always best.

Part III
THE DESIGN BRIEF

1 Explanatory

The aim of the design brief for a package is to bring together all those facts that the designer needs in order to carry out the job successfully. The brief will also serve as a blueprint for management.

In order to assist in the preparation of a brief, the pages that follow set out a series of headings covering all the marketing, functional and graphical aspects of a typical package. If those headings that are relevant to that package are then picked out, the fullest possible information should be set down against each one. This data will include physical information about the product, the results of research, management decisions about marketing policy and anything else that the designer needs to know and which management feels should be recorded.

The headings can be considered as a series of questions to which answers are required; some answers can be provided without any difficulty, others require research or have to be provided by management. The circumstances in each individual case will vary and some other headings – or questions – may have to be added. Because marketing is fundamental to package design, the headings have been grouped together in sections which are appropriate to the marketing plan. However, there are also a number of practical aspects which have to be included in the brief but which are only incidental to marketing.

The data required for the functional design of a package may have little relevance to the visual aspects, but both have to be considered together and both are relevant to the successful outcome. A visually satisfactory package which does not protect its contents, or an economical package which repels the customer, will add nothing to profits. All parts of package design are interdependent one with the other, and this must be recognised in preparing the brief.

Comments on the use of colour follow the discussion under each heading, where appropriate.

2   The product

2.1   Marketing aspects

The key to the whole design process is obviously the product to be packaged, because the nature and characteristics of the product dictate the nature and form of the package and affect the whole process of design. Before tackling the physical characteristics of the product, consider the marketing aspects.

The product   Define the product, and its colour, exactly. This may be thought rather superfluous, but much confusion can be caused if all concerned are not perfectly clear about the nature and form of the product which is to be packaged.

- Identify any colour particularly associated with the product, or which will enhance its appearance (see also 'Product associations'). Changes in preference for colour in relation to specific products may be significant; these may be dictated by natural progression or a desire for change.

The package   Define the function of the package in relation to the product:

The package may make the sale, as with a box of chocolates or a bottle of scent. A high quality package is prescribed.

The package may simply be an adjunct to sales, as with peas or jelly babies. The package can be utilitarian.

The package may be an essential part of the proposition, as with convenience foods. The type of package is important.

The form of the package may be dictated by the nature of the product as with frozen foods. A technical point.

The product may not exist without the package, as with aerosol paints. The package comes before the product.
It may not be practicable to distribute the product without a package, as with carbonated drinks.

- It is unlikely that colour will play any significant part here, but consider the point.

The nature of the sale The product may be an impulse buy or a considered purchase and this affects the nature of the graphic design. It may also affect the size of the package. A bar of chocolate is an impulse buy; a tea set is a considered purchase.

- A product bought on impulse requires a package in bright, strong colours, having impulse attraction. The colour of the package is a vital factor in creating an impulse sale; colours having maximum impulse attraction will invite the passer by to pick up the package, having been attracted by the colour. If purchase is a considered decision, more restrained colours are indicated and more attention should be paid to fashion and trends.

Sales characteristics These affect the nature of the package, for example whether the product sells singly, in a set, in bulk, in small quantities, for use in the handbag, as an accessory to something else, and so on.

- Colour selection is little affected.

Selling features The principal selling features of the product will need emphasis in graphic design. These include taste, flavour, nourishment, ability to wash whiter than white, and so forth. The selling feature may be an inherent property of the product such as reliability, value, newness and so on (see also Section 11.5 in this part).

- The principal selling feature should be reflected in the colours used, and hues with the appropriate attributes should be selected. The odour of the product may be particularly significant; a pine disinfectant will look best in a dark green bottle. If newness is a feature, particularly up to date colours are indicated.

Volume of sales The volume of sales expected will affect the type of package; high volume sales require a highly automated packaging line.

- No connotation for colour selection.

Repurchase If there are any special arguments in favour of re-purchase, such as refills, this will have an effect on the type of package and on graphic design.

- Although there should be a family likeness between original package and refills, the latter may be in different colours to prevent confusion.

Price If the product is low priced, this may be reflected in functional and graphic design, but an expensive product requires an expensive package made from good quality materials and having a sophisticated visual effect. Price may also place limitations on the cost of packaging.

- Price may have some effect on colour selection. A cheaper quality product would generally have brighter colours; the more expensive the product, the more sophisticated the shade.

Product associations Some products are associated with a particular type of package, for example most food packages are not suitable for disinfectants. Other things being equal, such associations should be followed.

- The colour of the package must associate with the colour of the product contained in it. A packet of peas would look odd if there were no green in the label, and it should be the shade of green that most people associate with fresh peas; the wrong shade might suggest artificial colouring, too bright a green might suggest preservatives. Consumers often associate colour with products; they have been found to prefer dairy products in blue and white packages. There may be some cases where it would be a mistake to match package colours to content colour too exactly. It is the colour that people associate with the product that is the important point to establish.

Illustration Decide whether the product is to be illustrated on the package.

- For colour selection, see Section 18.1 in this part.

Product ranges If a product is produced in more than one variety, flavour and so on, special consideration is necessary.

- For colour selection, see Section 18.2 in this part.

2.2 Product image

Product identity

The package ought to be a succinct impression of the image of the product that the manufacturer desires to put over, and the package may have to integrate with the product function. Establishing the identity of a product in the minds of customers is an essential part of good marketing and selling. Many manufacturers pay too little attention to product definition and the result is the confusion of customers. The package plays an important part in product identity because it is the essential link between manufacturer, retailer and customer. It is an integral part of the marketing plan and should reflect the promotional theme that it is desired to convey. The package should

- draw attention to itself and the product it contains
- act as reminder advertising of the product
- create an impulse to buy the product, because it has been advertised
- identify the product, the manufacturer and the brand
- promote the brand image where appropriate.

One of the aims of graphic design is to catch the eye of the customer, but it should also draw attention to the product. If the latter depends for its appeal on convenience, quality or some other attribute and this is promoted in press advertising, then the package must reflect those attributes both functionally and visually. If customers cannot identify the product when it is under their noses at point of sale, then the promotion is wasted. The most exhaustive study of the psychology of the consumer will be of no avail if the customer cannot recognise the product. It is never a good idea to try to attract attention at the expense of product identity, and the package must achieve product recognition and shelf impact; it may also have to achieve brand recognition.

The package should be a continuing reminder of the advertising of the product, and what appears in promotion should be easily recognisable at point of sale. Good design is the most cost effective medium in product promotion because it goes on working long after the campaign is finished and, in some cases, may be the only advertising medium for long periods. The best advertising ideas are those that cannot be separated from the products that they seek to enhance.

This does not mean to say that customers will always buy the product because they recognise the package. However, if they have seen it on, say, television, recognition at point of sale may create an impulse to buy. The package is particularly important to products advertised on television and, if possible, the design should be equally distinctive in colour and in black and white. The package should certainly have good recognition value; the eye of the purchaser only rests on a given spot for a fleeting second under normal retail conditions and if the package cannot be identified immediately it will be missed. No product will sell as the result of promotion unless the package forces itself on the attention of the consumer.

The principle also works in the reverse direction. Consumers are undoubtedly influenced by the large sums spent on promotion but the expenditure can be largely wasted unless the shopper is reminded of it by seeing the package at point of sale. A manufacturer advertises a product in the hope that people will seek it out, but consumers will be helped to recall the promotion by seeing the package.

Colour

To establish product identity it would be desirable to use colours which are associated with the product and which also suggest its most important attributes, particularly its principal theme of promotion. Even where colours are selected for other reasons it would be wise to consider these points. Colour can play an important part in creating a visual image of a product.

Corporate identity

The package should also be a visual expression of the corporate identity or personality of the manufacturer because this builds confidence in the product. It is particularly important with 'own label' products or where a manufacturer produces a range of broadly similar products, such as spices.

Package design is part of the visual image of a company, especially where the design incorporates a corporate symbol or logo; people do not always read names but they do recognise a visual pattern and consequently the symbol should be easily recognised and recalled. In many cases it will help to ensure maximum concentration on the package.

When two or more products are equally well promoted and customers are offered a choice at point of sale, they may choose the product with the most attractive package but they are just as likely to choose the products of the manufacturer whose corporate identity they recognise. Of all the competing products that appear in a selling outlet, only one will attract the purchaser at each sale, and the art of marketing is to ensure that sale by seeing that the package is both attractive and reflects the image of the manufacturer. It follows that it is important to study the packaging of the competition to make sure that these conditions are fulfilled; there must be sufficient distinction from the competition.

Laboratory analysis can reveal that the product differs from its competition but the consumer does not have access to a laboratory and may be confused by conflicting claims. The package design has to separate the product from its competition and communicate a positive message about the character of the product and the reputation of the company behind it.

A good package is a direct means of communication between supplier and customer and should convey to the customers what the supplier thinks of them and how the supplier has tried to meet their needs. The package should be unique, it should have a definite personality and it should convey a feeling that the supplier has the best interests of the customer at heart. The design should be clear, simple and an expression of the desired image. Too often designs are a conglomeration of colour, types styles and logos because the designer has tried to feature everything.

Colour

Colour can play a vital part in creating a visual image of a company, particularly where the design includes a trade mark or logo. A corporate identity may derive from colour, or from a combination of colours, perhaps in the form of a symbol or logo, but it would not usually be desirable to use the same colours for a whole range of products except in the case of generics. The logo might appear on the package in colour but the main colour of the package should have an association with the product.

Colour for own brand labels requires a good deal of careful planning because the design of the label should be such that it builds up an image of the store and creates consumer loyalty. The same label may be used for many different kinds of product and it may be unsatisfactory to use the same colour for all products. Coloured panels may be the answer.

Brand image

The package should reflect the brand image of the product or company. At this point it may be useful to draw a distinction between product identity, corporate identity, product brand image and corporate brand image. For example, confectionery usually has a product brand image whereas frozen foods have a corporate brand image. Either a product, or a corporate, brand image may be based on a symbol, trade mark or logo and this must appear on the package so that the customer recognises it; there may have to be a relationship with other products from the same source.

An individual product may have a brand image. Recognition of the image may depend to a great extent on colour because consumers prefer not to use their glasses when shopping and may not be able to read lettering. One of the best examples of this is Cadbury's dairy milk chocolate; they have used purple for their chocolate bars for many years, and the colour has become well established even though it is not particularly good in psychological terms. Rowntree have used black for their boxed chocolates in a similar way. Both are examples of consistent presentation.

In many cases the package is the brand and is its most effective form of promotion; in such cases it is the package that sells rather than the product. Advertising campaigns have been built on packages; the best known example is Benson & Hedges cigarettes, in whose advertising the product is not even mentioned. That packaging is an important element in creating a brand image is especially true of cigarettes where there is little difference between one product and another.

There may be some difficulty in identifying a product brand image by colour alone in self-service conditions because lighting may vary and the package may be displayed alongside a mass of competing packages using similar colours. Furthermore, colours do not necessarily look the same in different materials, such as on uncoated board as compared with coated board.

A company may also have a brand image which is extended to all its products, including packaging, and this assists promotion of the brand image. Promotion in one direction helps all promotion. Good design is very cost effective in brand promotion, but remember that the package must achieve brand recognition *and* shelf impact.

Colour

Either a product brand image or a corporate brand image may be based on colour, or on a combination of colours. If it is, the colour should be used consistently. The image seen in promotion, particularly on television, should be instantly recognisable at point of sale.

The selection of colours for brand images is a complex subject which is beyond the scope of this guide.

The remarks about colour under the heading of product identity (above) apply equally to product brand image but, where colour forms an integral part of either a product or a corporate brand image, do not sacrifice this if it can be avoided.

2.3 Physical characteristics

Physical form   Whether solid, liquid, powder, and so on. This dictates the nature of the package and whether a bottle, sachet, carton or other container is required. It also dictates the nature of the material that can be used.

- Only affects colour in a very general sense.

Physical characteristics The product may be hygroscopic, breakable, need protection against damp, be liable to pick up odours, and so on. The consistency of the product may also be significant; so may viscosity and flash point. These may affect the nature of the package and the materials that can be used.

- No significance so far as colour is concerned.

Wrapping The nature of the product may be such that it has to be wrapped individually or the container may have to have, say, a greaseproof liner.

- For colour selection, see Section 19.3 in this part.

Emptying The way that the product is taken out of the package may affect the design. An easy pouring device may be required and can be a useful selling point, or the package may have to dispense one pill at a time.

- Little effect on colour selection.

Adverse characteristics Products that are corrosive, chemically active, poisonous or sticky may require appropriate packages which eliminate breakage and guard against damage to the customer. Special labelling may be required.

- Adverse characteristics usually require special labelling and this may affect selection of colour. Labels must stand out.

Temperature range If the product is affected by, or likely to be subject to, extremes of temperature, special packages may be required. This applies equally to heat and cold.

- Bright colours on the outside of a package will reflect heat away from the container and this keeps the contents cooler; aluminium foil is often used for this purpose. Dark colours will absorb heat.

Evaporation A product that is subject to evaporation, or which may lose volatile elements, requires an appropriate airtight package and closure.

- No effect on colour selection.

Protection The product may require protection against breakage, tampering, pilfering, contamination, chemical interaction, rodents, insects and other hazards. The outer must also protect the contents.

- No effect on colour selection.

Rancidity   Special steps may be necessary to protect a product which is liable to go rancid.

- Colour is often used to protect against rancidity; film of an appropriate colour will be effective.

Toxicity   Careful investigation may be necessary to avoid toxic risks from packaging; this applies particularly to food packages and may place restrictions on the materials used.

- This is a technical problem and care must be taken that any colours which come into contact with the product are not toxic.

Wrinkling   Special types of package – and wrappers – may be necessary to prevent product wrinkling; this applies particularly to textiles.

- No effect on colour selection.

Processing   The product may require processing in the package, for example sterilisation, freezing, vacuumisation, refrigeration; this will affect the type of package and design generally.

- If the product has to be processed, the inks used for the package must be proof against the processing; this may place restrictions on the colours used.

Light   Some products require protection against light, not least against direct sunlight.

- Colour may be used for this purpose. Beer bottles are commonly brown for this reason and medicinal products are packed in amber coloured containers; the degree of protection is dependent on the wavelength of the light excluded by the colour. Transparent film used for wrapping may cause the colour of the product to fade; tinted film may be necessary to reduce the transmission of ultra violet light. Where the product is in a transparent container, some hues are more suitable than others for protecting the contents.

Spillage   Some products tend to disfigure the package when poured or spilled; this may affect design.

- Not significant for colour selection.

Safety   All packages should be safe during handling and in the home; they must be proof against children, safe to open, and safe to store. This point has to be considered in relation to the nature of the product.

- Colour is used to draw attention to hazards and to cautionary notices and instructions; red is commonly used to denote poison.

Product appearance   The package should enhance the appearance of the product contained in it. See Section 17.3 in this part.

- The colour of the package should not clash with, or detract from, the appearance of the product.

2.4   Types of packaging

Choice of types of packaging is a little beyond the scope of this guide because it depends on many factors which are the province of the technically qualified packaging expert. Nevertheless, choice of type of package is part of the design process and requires a few words.

There is a battle for supremacy between different forms of packaging which tends to fluctuate according to economic conditions. The demand for packages made from paper and board has been tending to fall because of rising costs due to adverse exchange rates. The board industry has made efforts to improve the appearance of cases, the main innovation being improvements in the appearance of outers by flexographic printing which prints the outer before it is stuck to the fluting; this makes the case attractive enough for supermarket shelves. The humble bag is holding its own and so are paper sacks; fruiterers and confectioners prefer paper but larger bags are cheaper in plastics.

The packaging industry is the biggest single user of plastics and, although plastics are considered to be a replacement for other materials, they have created markets of their own. The demand for plastics containers is steady and they are tending to replace the tin can which has been hit by rising costs. A recent development has been a plastics alternative to the tin can which can be used for heat treated and pasteurised foods. Flexible packaging is not used to its fullest extent; graphics can be made to sell harder in flexible materials. There is said to be a growing trend towards flexible containers which use a variety of materials. Butchers prefer plastics.

The glass industry is holding its own through the production of smaller, lighter and stronger bottles. Glass remains dominant in baby foods and coffee but not necessarily for other powdered materials.

The tin can has been hard hit by frozen foods and other changes in social behaviour; pet foods are the largest single user. Tinplate is still favoured for food cans but canned food sales are falling. Almost half of all drink cans are made from aluminium. Developments include enhanced decoration and a gloss finish which feels like enamel.

New developments in materials and types of packaging should be kept under constant review; a new development may be just what is required to provide a new product with that added something which ensures success.

Some recent developments that are significant include:

- The trend to convenience foods has created a demand for specialised types of package, including oven usable containers.
- The trend to superstores means increasing demand for higher quality individual packs, but less demand for outers.
- Heat process packs are popular with the public.
- One trip shopping encourages the use of PET bottles; consumers and retailers like them, but not designers. The base can be coloured.
- Shrink wrapping is increasing in popularity.
- Multilayer plastics containers have been developed for oxygen sensitive foods such as tomato ketchup.
- A plastics can for motor oil was developed as the result of a design brief by Mobil.
- Sales of containers of liquid soap, with dispensers, are a sign of changing social habits.
- The vogue for take away meals has given a fillip to aluminium foil containers.
- Aseptic packaging systems, including bag in box, are popular with consumers.
- Oxygen absorbing chemicals in the package can add more shelf life to snack foods and the like.
- Colour coded handigrip bags have been developed for oven ready chickens.

Colour

The type of packaging, and the materials used, will obviously have an effect on selection of colour because some colours cannot be produced in certain materials, or because the nature of the material places restrictions on the method of reproduction that can be used.

It should also be borne in mind that a given colour may look quite different on two different materials, for example uncoated board as compared with coated board. It is also important that the material should be exactly the same throughout a run; if any change is made, the formulation of the ink may have to be changed. This is particularly important when colour is standardised, such as where it represents a brand image.

Attention is drawn to the notes on plastics bottles in Section 19.2 of this part.

2.5 Package size

The size of package that is put on sale is primarily a marketing decision. However, it is a more complex problem than it appears at first sight and it must be considered as part of the design process, if only because a graphic design for one size may not fit another size. Size must therefore be considered at the outset and account taken of variations likely to be needed. Size is less a question of the actual weight of product included in the package but rather whether the requirement is for small packages, large packs, giant packs, multipacks and so on.

Shape and size may be dictated by marketing considerations, the needs of the retailer, display, handling or plant available. All the channels of distribution may have ideas on the subject, including wholesalers, retailers, supermarkets – and the consumer. The primary question is what the consumer wants, but this is not always very clear cut. Package size is often a habit; tea, for example, is commonly sold in 125g packs and customers would not take kindly to anything else, although they will buy tea bags in larger quantities. Conditions in the market do change and a change of size may suit the customer much better. Consider the following points:

Economy Customers may prefer small packs; or large packs at a cheaper price.

Economic conditions In times of lower spending power there may be a tendency to buy products in smaller quantities; slimmer forms of packaging may be indicated.

Frequency of purchase Weekly supermarket shopping predisposes to larger packages and supermarkets report that presentation in multipacks persuades customers to take more.

Ease of storing Today's kitchens are smaller and there is less room for storage; people buy smaller packs more often.

Ease of carrying Smaller packs are best for the corner shop but large packs for supermarkets where people have cars.

Ease of use Difficult to quantify but is always an important point. For example, portion packs may be useful in these days of convenience foods.

Freezers   The growing use of freezers may dictate larger packs or it may dictate smaller ones, depending on the product and the way that it is used in the home.

The outlets through which the product is sold will also influence discussions about package sizes. For example:

- Independent stores prefer smaller packages.
- Supermarkets and hypermarkets have a demand for large packs.
- Counter displays indicate smaller packs.
- All retailers want sizes that are easy to handle.

Where the product is or may be an impulse purchase, it may be a good idea to have smaller packs which will encourage impulse buying; people will buy a bar of chocolate on impulse but probably not a large box.

There are also internal factors that have a bearing on decisions about size. For example:

Cost   Smaller sizes are relatively expensive and the price of the smallest quantity may be dictated by the cost of packing. However, small sizes may be essential for sales reasons.

Plant available   It may be too expensive to change if new machinery is required.

Competition   Obviously the sizes offered by competition are significant.

Handling   Whatever sizes are used they must be easy to handle within existing constraints.

Outers   The package may have to fit into a standard carton or outer.

Value   A small package may create an impression of high value or exclusivity; colours should be appropriate.

Shelf space   The package should not take up more space on the retail shelf than its turnover warrants; design with store modules in mind.

Colour

There is little relationship between package size and colour except that different colours may be used for packages of different sizes, perhaps to encourage purchase of large sizes, or to give an impression of exclusivity to a small package. Colour can also be used to create an impression that a package is larger or smaller, heavier or lighter, and this may be useful in certain instances.

3   The market

3.1   Nature of the market

The product may be for sale in the mass market or at the top end. The mass market needs a less sophisticated approach to design, especially graphic design. The package aimed at the top end of the market may need snob appeal.

- The mass market likes bright and simple colours. The top 10 per cent wants something more sophisticated and will be attracted by 'different' colours.

3.2   Most profitable market

With a new product it may be necessary to decide whether to go for a universal market straight away or whether to aim at the premium section of the market. The volume market is not necessarily the most profitable.

- To reach a higher grade market, use colours that are different to those used in the supermarket; use more sophisticated variations.

3.3   Type of market

The product may be aimed at a convenience market or at the take home trade and the package must be satisfactory for that market.

- Unlikely to affect colour, but consider.

3.4   Speciality markets

The product may have an appeal to a sectional market and it might be
clothed in a special package for that market, or for a special occasion
such as Christmas.

- Seasonal markets require special colour treatment and appropriate hues should be used. A high fashion market requires colours that convey high fashion and snob appeal; this indicates
  sophisticated colours.

3.5   Export markets

If the product is to be sold in overseas markets, different packages may
be required.

- For colour selection, see Section 16 in this part.

4   The customer

4.1   Type of customer

If the package is aimed at a specific type of customer, then the design must reflect the characteristics of that type. Some types of package appeal to one type of customer, but not to others. Most packages are aimed at consumers but some may be aimed at business men or commercial purchasers.

- If the package is aimed at a specific type of customer, select colours that will appeal to that type. Consumers, for example, have different tastes to business men; the latter will react to hues that suggest tradition and respectability. The more highly educated a person, the more subdued colours are preferred. Commercial markets require colours that have a utilitarian purpose.

4.2   Age of customer

If the product or package is aimed at children, the package should appeal to them rather than to sophisticated adults. An out of date design may suggest inefficiency to the young, a modern design may not inspire confidence in older people. Special packs are sometimes designed to attract the young.

- Colour is very important. If the product is aimed at the young, select colours that will appeal to them – generally bolder and brighter colours like red, yellow and strong blue. Young people like large areas of white and bright colours; pastels take second place.

4.3   Sex of customer

A package may suggest feminine charm or masculine power depending on the aim of the product.

- A package that is bought primarily by women should be in feminine colours. A deodorant aimed at men might be in bolder, or more aggressive colours. See also Section 11.8 in this part.

4.4   Purchasing influence

The person who actually buys the product may be significant. Many products are bought by women for men and vice versa; women often buy products that they think that men would like. In some cases, husbands and wives are joint purchasers. Some products are bought by men for women. The purchasing influence may not, of course, be either the wife or the husband.

- Where a product is bought by women for men, they choose colours that they think men will like; most men's toiletries are bought by women but they are packaged in colours that appeal to men.

5 Purchasing habits

5.1 Buying habits

The buying habits of the customer can have a significant effect on the whole concept; purchasers may prefer to buy in large quantities, in small quantities, or in some special way. Customers may be conditioned to a certain type of package or to a standard unit of packaging.

- General buying habits should be considered; they may have some effect on colour.

5.2 Presentation

Purchasers may expect to see the product and to be able to judge its quality; a transparent pack may be indicated because the purchaser buys the product and not the package.

- Little specific effect on colour selection.

5.3 Reasons for purchase

The consumer buying food for the home will not be attracted by the same type of package as the same consumer buying chocolates for the theatre. The graphic design may have to vary in different markets. For example, Ovaltine is considered a bed time drink in the UK but a morning drink overseas.

- May have an effect on colour selection.

5.4   Motivation of purchaser

The motivation may be pleasure, necessity or utility; this can be reflected in the type of package and the graphic design.

- Colour might be used to create an appropriate emotion, depending on the nature of the motivation.

5.5   Pattern of purchase

This can change; hypermarket shopping involves less frequent purchases and may require different packages or package sizes. Because of one-trip shopping, packages are tending to become larger and there may be demand for resealable packages. Presentation in multipacks will often persuade purchasers to take more.

- Little effect on colour selection.

5.6   Shape and size of package

This is a complex subject. See Section 2.5 in this part.

- Shape and size have an effect on colour selection in an abstract sense. Certain colours enhance certain shapes and the skilful use of colour can make a package look larger.

5.7   Method of choice

The way that customers make their choice may depend on the attraction of the package but may depend on other factors; customers may, for example, be conditioned to handy packs.

- Unlikely to have an effect on colour selection.

5.8   Circumstances of purchase

The same person may have different needs according to the timing of the purchase. The same product may be bought for utility or pleasure and different designs may be used for different outlets.

- May have an effect on colour selection, depending on the circumstances.

5.9 Impulse purchase

Products normally bought on impulse require a particularly eye catching form of package. See Section 2.1 in this part.

- Use impulse colours.

6 Usage of the product

The way that a product and its package is used in the home may have a significant effect on package design.

6.1 The product

There may be a need for a package that will pour easily; a detergent pack will stand on the draining board and may have to withstand wet.

- Little effect on colour selection.

6.2 The package

The way that the customer can use the package may be a useful sales theme, such as coffee jars used for storage, but this can cause awkward production problems.

- The colour of the outside of the package may be important when the product is left in the package for long periods and possibly dispensed therefrom. A cosmetic package, for example, remains on the dressing table for long periods and should not only be attractive in itself but should also be finished in a colour suitable for use in the bedroom; the colour should also enhance the appearance of the product. In the case of cosmetics, female colours are indicated. If the package has a re-use value, consider colours suitable for the secondary use.

6.3 Protection

The package should not disintegrate before the contents are used up, and this may require special consideration for long lasting products.

- Little effect on colour selection.

6.4 Storage

A foodstuff that is stored in a refrigerator or freezer needs to have a package which is convenient for home appliances and it must, of course, stand up to freezing. Some products will be stored for long periods and the package needs to give long term protection.

- No colour connotations except that the colour should stand up to freezing.

6.5 Where used

A product used in the kitchen needs a pleasing design; the householder may tire of an ugly design and never buy the product again.

- A product used in the kitchen should be packaged in colours which do not look out of place in the average kitchen; it would be a good idea to follow kitchen trends. Similar remarks apply to bedroom products; pink would be very suitable for a cosmetic package or for facial tissues.

6.6 Suggestions for use

Recipes for cooking, methods of serving, hints on using and similar matters may be useful.

- No connotation for colour except legibility.

7 Selling conditions

The conditions under which the product is sold have a considerable bearing on both functional and graphic design of packages.

7.1 Type of outlet

This may be significant and it is necessary to identify the different needs of supermarkets, hypermarkets, department stores, variety chains, discount stores, ordinary retailers, mail order houses and others in packaging terms. What is good for a chain store is not necessarily good for an exclusive store, and a package that has to take its chance in a dump bin in the supermarket may not be suitable for a speciality store.

Specialised shops tend to have specialised needs; butchers are said to prefer plastics but fruiterers and greengrocers prefer paper.

Colour

The type of store will govern the method of display. In self-service stores, hypermarkets and the like, colours must have sufficient impact to attract the attention of the customer and ensure sales.

More sophisticated colours are required for packages which sell in higher grade or specialist stores where there is less necessity to catch the eye of the passer by.

Sameness of colour and typography can be a fault, especially where own brands are concerned; the use of blue by Boots and of orange by W. H. Smith does very little for sales.

7.2 Method of display

It cannot be assumed that every package will be displayed in the same way on a supermarket shelf. For example:

- A product may be sold direct from the outer and the latter should be part of the image; impulse colours are important.
- Some packages may be specifically designed for counter display, or check-out display.
- A package that is used for dispensing loose products needs to stand up to handling.

Colour

When the package is sold from an outer or display case, the colours must be arranged in such a way that the whole does not become a heterogeneous mass; units must not lose identity.

The normal background to display may be important if it can be pinpointed with accuracy. The colour of the package should not look out of place in normal conditions and should stand out against the background, but a too dominating colour may do more harm than good.

7.3 Shelf display

A large majority of packages are displayed on the shelf in self-service or similar conditions, and the package must be easy to stack, make the best use of shelf space and be easy to position so that the best selling face is to the fore. The package should be capable of being identified from every angle; for preference, the design should be repeated on every face but at least the name should be on each face.

The right spot on the supermarket shelf is a matter of life and death to some products and every merchandiser struggles to secure it. The right spot is said to be arm high for a woman of 5 ft 4 in, but some packages will be displayed at eye level, others will be on the bottom shelf and product information should be visible from all angles. The eye of the purchaser meets the shelf at an angle and cannot always perceive a clear cut front panel. Other points:

- Upright facings permit a greater number of facings per length of shelf but, if the packages are displayed horizontally, the design must be modified accordingly.
- Packages displayed on end require special consideration.
- The product information should be clearly visible when the package is displayed to best advantage.
- A package that is too dominant may be monotonous and enhance

the appearance of competing packs; it may invite the customer to try something else. Some packages, however, *ought* to be aggressive and dominant.

- Visibility will depend on the conditions under which the package is seen; some may have to be seen in dark basements.

Colour

A package that has to take its place on the supermarket shelf needs display colours; red is very suitable, but avoid sameness of colour. If everybody follows the same rules, every package will look alike and it may be a good idea to be 'different'. Colour may be used in such a way that the package looks its best when stacked on its shorter side. Colour must not hide the product information.

7.4   Lighting

Designs can look quite different under various forms of lighting; design for the conditions that are normal at point of sale. A package which is consistently seen under 'daylight' type fluorescent lighting needs to be tested for effect.

Colour

Colours should look at their best under normal shop lighting but it should be remembered that package colours will look different in daylight, fluorescent lighting and the normal tungsten lighting of the home. It has been found that some combinations of blue and white look very insipid under supermarket lighting.

8 Distribution conditions

It is all very well designing a package which has maximum effect at point of sale, but it is often forgotten that the package has to reach the point of sale in good condition and the nature of the channels of distribution may influence design. Too many firms design their packages to help themselves and not their customers; their customer of the first instance is the retailer and the package has to reach the retailer safely.

8.1 Channels of distribution

The channels may include a wholesaler as well as the retailer; the product may be delivered to a chain store warehouse; or the product may be delivered direct to the supermarket. All these channels affect the nature and design of the package and more particularly the outer or container. The wholesaler or warehouse needs to be able to break bulk easily; the retailer is primarily concerned with shelf filling and pricing. The supermarket may dictate the form of packaging they will accept.

- Little effect on colour selection.

8.2 Storage conditions

These may influence the shape and size of the package and the way that it is packed into outers. The packages, or the outers, may have to be palletised or placed in cages.

- Storage conditions may be significant for colour. Some colours fade away under dim lighting and this may affect the reading of labels or other information.

8.3   Case size

Most products are sold by the case and the size of the case needs to be thought through; a case containing a dozen of some products might be too heavy, or inconvenient.

- Little effect on colour selection.

8.4   Information

Cases and outers need to have visible product information which can be easily perceived when stacked in a warehouse or storeroom.

- Colour is important from a legibility point of view.

8.5   Pricing

Packages should be capable of being priced easily while still in the outer, or on the shelf. A price spot on the package may be useful.

- Little effect on colour selection.

9 The retailer

9.1 Retailer needs

The power of the retailer is an increasingly important factor in market planning and manufacturers have to align their ideas to the considerable power now exercised by the retailer, especially the supermarket and hypermarket chains. The larger multiples are in a position to dictate the nature and shape of a package and the way that it is handled. The management planning a new package would be well advised to take account of this trend and to seek co-operation with major retailers before finalising package design. The indication is that more direct sell packages will be required and that cage pallets will be increasingly used between factory and supermarkets.

This is not the place to discuss the philosophy of retail selling, or the significance of own brands or generics, except to say that own brands now account for about 25 per cent of sales through multiples and this puts pressure on manufacturers who are promoting their own brands. Some sources say that own brands are not doing so well, but this depends on the retailer. Tesco, for example, say that they are tending to be more and more innovative because manufacturers are not giving them what they need. Own labels are said to be likely to exceed generics because the former provide a higher profit margin and because own labels have to be packaged just as much as proprietary brands. This seems to suggest that good packaging is just as important, whatever the nature of the trade.

Asda are said to prefer proprietary brands but they do expect the manufacturer to respect their skills in marketing. Asda do not like the package that will not sell the product but they do stock products that they know would sell better if they were properly packed; they also say that many non-food items are quite unsuited to superstore trading because the packaging is unsuitable. Packaging improvements can play a major part in increasing volume sales in superstores.

Most retailers are unlikely to allocate display space unless they are convinced that the product will sell. Therefore they are interested in the marketing aspect and expect the product to be supported by suitable promotion so that purchasers are persuaded to come into the store and seek out the product. The relationship between advertising and packaging is important to the retailer.

Retailers will look most kindly on those packages which require the minimum of handling to put on the shelf and which take up least space in relation to turnover; the criterion used by most supermarkets is the rate of sale per square metre. An awkward shape may take up far too much room on the shelf, as well as being difficult to stack. It should also be easy to price and, if the customer can carry it home without further wrapping, so much the better.

As far as the retailer is concerned, it is the practical design of the package which is important and the view of the retailer at an early stage in the design process would be useful. The visual attraction of the package plays some part in the decision of the retailer but this is only incidental to other aspects.

- The views of the retailer are not relevant to colour selection.

9.2 Size and shape

The package should not take up more space on the shelf than its turnover warrants. The package should be designed with the store module in mind, especially very small items. See also Section 2.5 in this part.

- No effect on colour selection.

9.3 Regulations

The package must conform with labelling and other regulations; this is the responsibility of the manufacturer and the retailer is entitled to look to him for protection. Bar coding should not be forgotten. See also Section 15 in this part.

- No effect on colour selection except that labels must be legible.

9.4 Protection

The package must protect the contents under normal retail conditions; no retailer wants damaged goods or wastage.

- Not applicable for colour selection.

9.5 Dating

Consider the need for 'sell by' dates where applicable.

- No effect on colour selection.

9.6 Up to date

The package should not 'date' in design terms, or give any impression that the product is old fashioned. Nor should radical changes be made which will cause old stock to become obsolete.

- Up to date colours, or variations of colour, are indicated.

9.7 Image

The package should link with the image of the manufacturer put over in promotion.

- For colour selection, see Section 11 in this part.

9.8 Display

The package should lend itself to display and should add to the appearance of the store; it should invite attention from customers who come in for something else.

- Colours should not be so startling that they look out of place in a well run store.

9.8 Advertising

The package should link with advertising so that it is immediately recognisable on the shelf.

- For colour selection, see Section 11 in this part.

9.10 Pre-pricing

Some supermarkets believe that pre-pricing is valuable and adds to the clarity and design appeal of the package.

- No effect on colour selection.

10 Competition

10.1 What is the competition?

Any package must be different to, and better than, the packages of competing brands which may appear beside it on the shelf. The package should stand out from competing brands; it will not sell if it is lost in dozens of competing packages. Knowledge of competing products, competing brands, and their packaging methods is therefore essential and what the competition does should influence design.

The competition may not necessarily be another brand; it may be another product, for example margarine is competitive with butter.

- Colour may have to be different to competing products.

10.2 Who is the competition?

This includes competing brands and competing products whose packages are likely to be on the same shelf.

- Colour should be different from competing brands.

10.3 What does the competition offer?

The method of packaging and the type of package used by the competition should be analysed.

- No significance for colour selection.

10.4   Advantages

What advantages does the product have over the competition? These should be emphasised in design.

- Colour may be used to underline the advantages where appropriate.

10.5   Distribution

Consider the distribution methods of the competition and their implications.

- No significance for colour selection.

11 Method of promotion

11.1 Advertising generally

The method of promotion, and the theme of promotion, need to be carefully considered in the approach to package design so that the two may be co-ordinated. There are three main factors involved:

- the message that it is desired to communicate
- the selling features to be emphasised in promotion
- the type of promotion, and co-ordination of package and promotion.

Asda has pointed out that there is insufficient co-ordination between promotional ideas and package design and that design planning often neglects the ways in which the product is to be promoted. Promotion takes a number of forms and may reflect a specific sales theme or may be based primarily on promoting the brand. The points that need to be considered at the design planning stage are:

- press advertising
- television advertising
- special offers
- brand promotion
- specific sales themes
- seasonal sales
- gift sales
- mood
- the package itself.

The design of packaging tends to come before the formulation of an advertising campaign, and consequently the package design often does not co-ordinate with the advertising and has little relevance to the image created by the advertising. There should be a relationship between the package and the advertising of the product, especially when television advertising is used. When the customers see the package on the shelf, they should be able to recognise what they have seen on the screen or in the press, and to relate the two. The package should have a clear identity which carries through all forms of promotion and it should mirror all the main ingredients of the marketing mix, including the concept behind the product. The package should be an extension of product advertising and both shape and graphic design may be involved. However, the package should also be complete in itself, because the package may be seen but not the advertising; in any case, it is unwise to tie the package too closely to the advertising because the package is still there when the advertising campaign has been forgotten. It may not always be possible to relate the graphics on the package to the advertising and it is seldom a good idea to use advertising copy on the package; the copy may not succeed, although the package does.

- The colour of the package should co-ordinate with the advertising of the product. If the advertising is black and white, the graphic design is relatively more important than colour.

11.2 Television advertising

It is important that the image shown on the screen should be easily recognisable at point of sale, or the whole operation is wasted. The package seen on the shelf should remind customers of what they have seen on the screen, and this helps to create a sale. Wording may be of little significance without shape, and visual impact is vital.

- Special care with colour is necessary if the package is to appear on television because some colours do not reproduce well on the screen. Remember also that not everyone has colour television – black and white is still significant.

11.3 Special offers

The design of the package may have to lend itself to various methods of promotion such as free offers, money off, coupons and the like, and space may have to be left for the special offer. Such packages are often designed specially for the promotion.

- Superimposed colours for special offers should contrast pleasantly with the normal package colour, but impulse colours should generally be used.

11.4 Brand promotion

Packaging is an important element in the creation of a brand image. This is particularly true of cigarette advertising where there is little apparent difference between one brand and another. Attention is drawn to Section 2.2 in this part.

- Colour as a means of creating brand identity or brand image requires special care.

11.5 Special sales themes

In the package design context this includes those features of the product which may be emphasised in promotion. Examples are convenience, handy for the pocket, easy to carry and so forth; a large and bulky package would do nothing for a product which was promoted as being easy to carry in the pocket. The sales theme also includes, of course, features such as nourishment, value for money, taste and so on.

The sales theme should be considered in the conception of the package and may include both functional and graphic design. For example a product sold with a convenience theme benefits from an appropriate package such as 'boil in the bag'.

- Colour may be selected to support specific sales themes. See Section 2.1 in this part.

11.6 Seasonal sales

Many products have marked seasonal sales and packages may be designed specifically for various seasons. Special Easter packages or Christmas packages and wrappers are typical examples.

- Colours appropriate to the seasons are:

  spring   bright, pale colours, especially yellow and green.
  summer   darker shades than for spring but suggesting sunshine and light.
  autumn   browns and beiges.
  winter   brown and other warm colours; red is appropriate for Christmas.

See also gift sales in the following section, and also seasonal wrappers, discussed in Section 19.3 of this part.

11.7 Gift sales

If products are purchased primarily as gifts, or if gift sales are promoted, the package may have to be designed to appeal to those likely to buy gifts, such as the young man buying a present for his girl friend, or the husband taking a present to his wife. Seasonal gift sales are also a possibility, for example Christmas, Easter.

- Special gift packaging may be devised in appropriate cases; colours should usually be bright and cheerful or appropriate to the season.

11.8 Mood

Depending on the product, and its sales theme, the marketing plan may call for the package to express a specific mood or emotion. This is a combination of shape, form and colour which is manipulated by the designer to convey the desired emotion.

For example, the marketing plan may call for a package with an austere or clinical look; or it may call for a friendly, exciting or comforting appearance. A convex shape, for example, is friendly and bright colours are exciting; a very vivid colour may excite curiosity. The package may have to convey a beneficial image, such as the nourishing qualities of a soup, or the safety qualities of a disinfectant. Nostalgia is an emotion that sells particularly well.

The package may also be required to have an appeal to one sex or the other. This is partly a function of shape and colour but may also be indicated by the design. For example, blue stripes, a piece of rope or a wooden surface are masculine in nature; soft pinks or velvet ribbons are feminine (see also Section 4.3 in this part).

Some shapes, colours and designs have unpleasant associations and should be avoided when known. For example, certain shades of yellow green tend to suggest sickness.

Texture may sometimes be used to create an appropriate atmosphere, although this depends on the nature of the material used for the package.

Any package ought to impart an impression of value for money.

- The colour of the package may be selected to express a mood or emotion such as nostalgia, excitement, curiosity, and so on.

11.9   The package itself

The nature of the package may be a selling feature in its own right, for example

- portion packs for food products
- carry home packs
- a package that can be reused
- unbreakable packages, where appropriate
- childproof packages, in appropriate cases
- packages that can be carried in the pocket.

Packages can also have adverse features. A package that is difficult to dispose of should perhaps be avoided. In certain cases the package itself may the feature of promotion (see Section 2.2 in this part).

- No special comments about colour are required. The colour should associate with the product or with the sales theme where appropriate.

12   Limitations on design

12.1   Machinery available

The ideal design from a marketing point of view may be limited by constraints imposed by existing machinery and other factors, and these have to be identified and assessed when planning a brief. Existing machinery may dictate the shape and size of the package; it may be too expensive to purchase new plant and the load on existing machinery may be a decisive factor in costs.

- The type of machinery may place limitations on the colour, or number of colours, that can be used.

12.2   Layout of the plant

This might have to be changed; a new type of package might require relocation of the production line.

- No effect on colour selection.

12.3   Availability of materials

The nature of the product may suggest the use of certain materials for the package but, if these are not easily available, alternatives may have to be found.

- The materials used may place restrictions on colour.

12.4   Economy of materials

The need for economy in materials may impose restrictions. Quantity may make some materials uneconomic or a design might be based on uneconomical board sizes.

- Economy may also impose limitations on the number of colours used, or the costs of some pigments or inks may be prohibitive.

12.5   Production tolerances

The package must work on the machine and it may be necessary to allow for creep and wander on the machine.

- Production tolerances may place limits on the amount of colour, or the way that colours are combined.

12.6   Different materials

If the package is composed of two or more different materials, there may be limitations on design because the materials require different production processes.

- Different materials may require separate methods of reproduction; this may impose limits on colour.

12.7   Cost

A penny on the cost of a package may make all the difference between profit and loss and the price that can be charged for the product may be governed by the competition; packages must be designed within constraints imposed by price. In some cases, such as perfumes, the cost of the package is the major factor in price.

- Similar remarks as for 'Economy of materials' above.

12.8   Wrapping

If the package has to be wrapped, and especially if the wrapping has to be heat sealed, this may impose limitations on both functional and graphic design.

- For colour selection, see Section 19.3 in this part.

12.9 Materials used

The materials may impose limitations on both production and design. For example, a blow moulded container in a difficult shape may slow the production process and be far too expensive.

- Materials used may impose limitations on colour for either technical or economic reasons.

12.10 Shape

The shape may impose limitations on the amount of decoration possible and on the reproduction processes used.

- Shape may impose limitations on colour for technical reasons.

13 Technical features

13.1 Type of closure

In designing any package there are a number of technical features which have to be considered when selecting the type of package; these may be dictated by marketing considerations or by the nature of the product. Some are only incidentally part of the appeal to the customer.

The type of closure may be dictated by marketing considerations, the nature of the product, trade custom, or availability of closures. The type used would obviously affect the design of the whole.

- For colour selection, see Sections 19.1 and 18.2 in this part.

13.2 Form, fill, seal

If this method of packaging is used it will restrict the size of the package and its nature and also affect the graphic design.

- This may affect the combinations of colours used and the way that colour is used, rather than the actual hues.

13.3 Sealing

A form of seal may be required which will protect against tampering, contamination, breakage and so on. Childproof seals may be obligatory in some instances.

- No effect on colour selection.

13.4   Reclosure

A foolproof recloseable seal may be required in some instances and may be a selling point.

- No connotations for colour selection.

13.5   Ease of opening

This is essential from the consumer angle and it should be possible without damaging the contents. An outer also needs to be easily opened without a tool kit.

- No connotations for colour selection.

13.6   Breaking bulk

If bulk has to be broken by the retailer, for example when selling by the ounce, an appropriate type of package is required.

- No connotations for colour selection.

13.7   Structural strength

The package must be strong enough for its intended usage, including usage in the home.

- No connotations for colour selection.

13.8   Mechanical filling

The package may have to be designed around a specific method of filling.

- This may affect colour combinations and the way that they are used.

13.9   Use life

The package should be able to stand up to wear during the life of the contents; this includes life in the home.

- No connotations for colour selection.

13.10   Shelf life

The package must, at least, protect the contents adequately until purchase.

- No connotations for colour selection.

13.11   Finish

Varnishing or lamination of a package makes for added appeal, especially where a high quality image is required. This also includes the use of coated board and other methods of obtaining a high gloss finish.

- The finishing process used may impose some limitations on colour and may also have an effect on the appearance of the colours.

14 Reproduction

14.1 The substrate

Those concerned with the design planning of packaging should be aware of the reproduction processes in common use because these often affect the nature of the design and may impose limitations on what can be done.

The nature of the product will usually dictate the nature of the material used for the package, and this in turn governs the reproduction processes that can be used.

- The nature of the substrate may restrict the colours that can be used and the appearance of the colours; for example, the same hue looks different on coated and uncoated board. The substrate should be exactly the same all through a run; if any change is made the formulation of the ink may have to be altered. This is *especially* significant when colour is used to represent a brand image.

14.2 Direct print

Decide whether the package is to be directly printed or whether a label is to be used.

- For colour selection, see Section 18.4 in this part.

14.3 Method

Consider the method of reproduction and whether the design is suitable for that method; some methods of printing are more expensive than others.

- The ability to reproduce colours must be considered.

14.4   Labels

If labels are to be used, consider the method of printing and the material; this may require modification of a design.

- Labels generally permit richer colours to be used.

14.5   Illustration

Pictures of an actual product may not reproduce satisfactorily; it may be better to use (for example) serving suggestions. Avoid misleading illustrations.

- For colour selection, see Section 18.1 in this part.

14.6   Colour

Consider the number of colours envisaged; special colours may have to be eliminated to keep down costs.

- Special colours, or difficult colours, should generally be avoided in the interests of economy.

14.7   Inks

Printing inks may have to stand up to processing or, for example, to cooling or freezing. Toxicity may be a problem.

- The nature of the inks may restrict the colours that can be used.

14.8   Shape

An awkward shaped package may be difficult to print and less costly alternatives may be required.

- No connotations for colour selection.

14.9 Preprinted containers

These can be a headache if there is a product change or a change in marketing policy.

- For colour selection, see Section 18.4 in this part.

14.10 Quality control

This is important; bad printing and faded inks may suggest a bad product. Bad printing or bad colour can damage a brand image.

- Good colour and adhesion to standards is essential. See also Section 14.1 in this part.

15 Labelling

15.1 Product regulations

In the present context labelling refers to the data which must be included on a package, and not to the significance of labels applied to the package as opposed to direct printing. Correct labelling is vital, not only to keep on the right side of the law, but also to provide information about the product and how to use it. It is not proposed to try to set out all the statutory requirements in this book, but these requirements may affect the nature and size of the graphic design. It is particularly important that labelling should be legible and easily read (see Section 18.3 in this part).

It is necessary to ascertain the labelling requirements applicable to the product contained in the package; food labelling regulations are particularly onerous and all mandatory information must be included.

- Consumer regulations do not, as a general rule, dictate colour for mandatory requirements, but the data must be legible and this may affect the background colour.

15.2 Weight

This should usually be shown whether regulations require it or not.

- No connotations for colour selection.

15.3 Hazards

Poisons must be clearly marked and any other hazards should be mentioned. There are specific regulations about the labelling of hazardous chemicals.

- There are many regulations on this point and standard colour combinations are specified for hazardous chemicals.

15.4 Storage

Instructions about storage may be required in many instances, including foodstuffs and especially frozen foods.

- Colour selection is only affected to the extent that instructions must be legible.

15.5 Cooking

Cooking instructions are increasingly important for foodstuffs; instructions for microwave cooking are becoming common.

- Colour selection is only affected to the extent that instructions must be legible.

15.6 Price

Incorporation of price on the label is desirable where practicable and provision may also have to be made for price spots.

- Preprinted prices should contrast with the background; price spots should generally be left uncoloured.

15.7 EEC directives

Those that cover labelling should be studied and followed.

- No connotation for colour selection.

15.8 Bar coding

This must be allowed for.

- Bar codes are usually printed in black and the background colour must be such that they are legible.

15.9   Product content

Although this must be shown in most cases by law, it is increasingly being made a promotional theme by underlining fat content, sugar free and so on. This needs consideration.

- Some suppliers are using different coloured labels to indicate the nature of the contents of food packages, for example those having low fat in one colour, sugar free foods in another colour and so on. This is chiefly promotional in nature.

15.10   Identification

Different flavours may be indicated by changing the colour of the label or the colour of the closure.

- For colour selection, see Sections 18.2, 19.1 in this part.

15.11   Instructions for use

These may be required in many instances and, if they cannot be provided on the package itself, a separate tag or insert may be required.

- Colour may be used to draw attention to the instructions and the result should be legible.

15.12   Package contents

In many cases it is desirable that the package should list its contents, particularly in those cases such as household appliances where accessories may be included in the same package as the main product.

- No special connotation for colour selection.

16 Export

When a product is sold in markets outside the UK, special consideration needs to be devoted to packaging construction, design and colour. The needs of export markets should be considered during the initial design planning, and sufficient time and effort must be devoted to developing suitable packaging. Adequate market research is necessary to establish the acceptable sizes, colours and other characteristics suitable for each market, not forgetting labelling and other regulations. The whole package concept may require rethinking because the sales appeal of the product may have different facets in different countries. An international package may, however, have advantages in many cases. For example, quality and flavour may be associated with the originating country; Scotch whisky is a typical example, and Baxter put a tartan label on their soup cans with advantage.

If at all possible, a package design should be suitable for both export markets and home markets; otherwise there may have to be short runs, especially for overseas markets. This may be unavoidable if the package has to have print in foreign languages, although it may be possible to minimise the problem by careful design. Illustrations which can be understood universally will avoid the necessity for print, and instructions for use in appropriate languages can be attached separately.

16.1 Designing for export

The following are design points that require special consideration when overseas markets are concerned. The discussion is additional to that in previous sections in this part (reference number in brackets after each heading).

Marketing aspects (2.1)

The product   Consider whether the product is right for markets outside the UK and whether it will travel satisfactorily. There are some products which are just not suitable for overseas markets.

Package function   The package must be suitable for the markets in which the product is sold or will be sold; a package suitable for the UK is not necessarily suitable for East Africa. Ideally, a package should be designed for each market but it may be possible to develop a common form of package which can be adapted for various markets. Some countries are suspicious of packaging and it might be better to sell loose.

Sales characteristics   The nature of trade may be different. In some countries the unit of sale is very small and gross lots would be quite unsuitable for the average trade.

Selling features   Some products are bought because they *are* foreign, such as French wines, and the package should emphasise the origin.

Illustration   Packages will usually benefit from illustration. See Section 18.1 in this part.

Product image (2.2)

The exporter should have an easily recognisable corporate identity. The brand name should be pronounceable in a local dialect and should not convey any confusing meaning; it may be an advantage to retain an English flavour. Once a design has become established in a market, it is very difficult to change it or ease it out.

Physical characteristics (2.3)

If the product is fragile, it may need cushioning against shocks, especially if shipped by sea.

If the product is light sensitive, special containers are prescribed. Shelf life may be much longer in overseas markets and require a different type of package.

Types of packaging (2.4)

Decide the most acceptable type of container for the market. It might be a metal box, or it might be a plastics bottle. It may be possible to use the same container for both home and overseas markets, but an export container will often be an unnecessary expense at home.

Package size (2.5)

Sizes of package, range of sizes and measures will be different in each market. The product may have to be sold in certain unit sizes in some countries.

The market (3)

Consider which overseas countries present a market for the product and, in each country, watch for changes in conditions and regulations.

Market customs may limit colour and design and certain forms may be unacceptable in some countries. International design does not always work; design may have to be aimed at a specific country because of climatic conditions, religious requirements, or because labelling regulations are different.

Foreign language equivalents may be required. It may be a good idea to separate label design from text; the latter can be altered more easily in language terms.

Purchasing habits (5)

The reason for buying the product may vary from country to country and require different types of label, or even different packages. Ovaltine is bought as a morning drink everywhere in Europe except in England, where it is a night time drink.

Selling conditions (7)

Some overseas countries have self-service stores but others do not, and the package might be lost in the gloomy interior of a bazaar. Display outers may be a waste of money in some overseas markets.

Distribution conditions (8)

What treatment will the package receive during transit? An extra strong package may be required in some markets.

Consider freight costs; avoid unnecessary packaging, but do not underpack. How many items to an outer are desirable?

The retailer (9)

In what form will the dealer want to receive the product?

Competition (10)

The competition, and its packages, will be different in each market.

Method of promotion (11)

In illiterate markets the package is the principal selling point, but it should usually have advertising support and there should be a close association between product, package and advertising. The sales theme may have to be put over in a different way in overseas markets and this may affect the design of the package.

Technical features (13)

Different packages may be required for tropical and temperate zones and in less developed countries the package comes in for more wear and tear. Export packages may require extra sealing to avoid pilfering and substitution.

Climatic and insect attack may have an effect on the product and/or the package and special precautions may be necessary. The package may be subject to extremes of temperature and may require a degree of insulation, or special adhesives. Humid conditions, or damp, may be an important factor and may make cartons unusable; if the product is hygroscopic, special precautions will be required.

Different markets may require different types of closure.

Bulk may have to be broken into very small units, perhaps on the dockside, and this will affect the packaging.

Labelling (15)

Packaging regulations vary from country to country, although the EEC is progressing towards standard European regulations.

Directions for use may have to be expressed in an idiom that is generally understood and which is not necessarily the local official language. Translation should always be good, and idiomatic. A stick-on label in local languages may solve some problems. Where descriptive matter is provided in a number of languages, no one language should be given priority; an insert may have to be different for each country.

Give the country of origin as 'Great Britain'; this is usually understood. Regulations may require colouring matter to be specified and the materials in the product must be acceptable in the countries concerned. The product specification, or production specification, may have to be shown on labels and materials may have to be stated in specific unit sizes.

Customer prejudices (17.1)

Local sayings, customs and superstitions may have an effect on design; colours and symbols associated with good luck are good, but taboos will kill sales.

16.2   Colour for export

The importance of the subject justifies extended notes on the subject of colour for export packages. Colours that are suitable for the British market are not necessarily suitable for overseas markets. Generally speaking, colours acceptable in British markets are also acceptable in northern European markets, but every country has its own associations, traditions and prejudices which can only be pinpointed by research in each individual territory.

Colour preferences are different in each country and are linked to levels of sophistication and also to climatic and racial conditions. Simple people generally like bold, strong colours and so do people who live under a strong sun because colour has to compete with the strength of the sunlight. People living in temperate climates react better to restrained colours.

Trends of consumer preference for colour also vary from country to country; there is comparatively little variation between trends in the UK and trends in northern Europe, but there are considerable differences between the UK, the USA, Canada and Australasia. Even in northern Europe there are often local differences relating to specific areas of the home which may affect packaging. For example, a range of kitchen colours suitable for the UK would not necessarily appeal to German purchasers. The position varies from time to time; UK and US trends are quite close at some periods but are far apart at other times.

Where a package is intended for use outside the UK, it would be wise to research each country in which it will be used in order to pinpoint specific prejudices and associations in that country. The symbolic use of colour, in particular, depends on the country. Colour stability and light fast qualities may be a problem in some countries and climates, and the nature of the climate may have an effect on colour appearance. The racial origin of purchasers may be significant in some instances, although climate is generally of more importance. Many of the prejudices, likes and dislikes that apply to individual countries are often a reflection of ethnic origins, although they may also be a reflection of local living and climatic conditions. If the package incorporates an illustration of people, it is usually a good idea to portray people, in the illustration, whose skin colour is the same as the inhabitants of the country concerned.

The following comments on individual countries have been noted during the course of research and derive from experience. However, the list is by no means exhaustive and it is strongly recommended that each country should be studied individually, and in detail, when export packaging is under consideration.

Africa generally   Be aware of differences between black and white cultures; a design which appeals to one may not appeal to the other.

Use bold colours.

Humans shown in illustrations should be healthy and strong and should usually be shown in Western dress.

Avoid the use of colour for identification, such as of flavour.

Avoid the use of peacocks as a symbol. Roosters are a symbol of virility.

If a well established product has to be altered, display the new package in white shops first, otherwise blacks may think that they are being offered something inferior. It may be necessary to consider after-use of packages; factors which appeal to blacks may not be understood by whites.

Symbols are often more important than colour, but as a general rule do not use animals as a symbol.

If colour is changed, leave the shape unchanged.

Arab countries   Take care when using green because it has Moslem connotations; avoid designs incorporating a cross. Avoid blue and white combinations; they are the colours of the Israeli flag.

Australia   Colour preferences are not the same as in the UK and a European colour range is unsuitable for the market. Australia is much nearer to the US, in colour terms, than it is to the UK.

Austria   There are sex limits to what can be shown in illustrations.

Buddhist countries   Saffron yellow indicates priests.

Canada   In general, colour preferences are similar to the United States.

China   White is the colour of mourning and white robed figures in illustrations might have an adverse effect. Red is a happy, profitable colour. Blue and white together mean money.

Czechoslovakia   Green means poison and a red triangle is also used to indicate poison.

The East generally   Yellow means plenty; yellow and pink together connote pornography.

Egypt   Black is associated with evil. Highly coloured complexions are prized.

Europe generally   The EEC is a collection of separate markets and not a European market, and what is acceptable in one country is not always acceptable in others. Swastika designs are disliked.

France   The French are not too concerned about colour but they do expect a reasonable package; design tends to be flowery, full of fantasy and sophisticated. Green is associated with cosmetics. Avoid illustrations showing liquor being poured.

Germany   Packages should be straight, clean and modernistic. The Germans have an ascetic sense of design; it should always be as simple as possible without unnecessary trimmings. The use of super-latives in advertising copy is forbidden. *Mist* is the word for dung; the word *Gift* means poison.

Hong Kong   White packaging is not acceptable and cats should not be illustrated on tins. Floral designs are safest.

India   Avoid using cows as symbols and avoid using skinny models in illustration.

Ireland   Green and orange should be used with care whenever there is an Irish element.

Israel   Avoid yellow because of its association with Germany.

Moslem countries   Green is a holy colour and should be used with care; most Moslem countries have green in their flags.

Netherlands   The Dutch prefer solid respectability to licence. Avoid using German national colours for packages.

Norway   Consumers welcome packages with plenty of colour.

Pakistan   Saffron and black are the colours of hell.

Saudi Arabia   The human form may not be used in advertising.

Spain   Bikini clad girls are best not used for illustrations.

Sweden   The Swedes are particular about packaging and do not like goods packed in gold or blue. Combinations of white and blue, the colours of the national flag, are best avoided. Regulations are more stringent than in most other countries and packages usually have to be designed specifically for the Swedish market. Consumers do not like giant packs. The trade name must be pronounceable in Swedish. A modern decor is preferred.

Switzerland   Yellow conveys cosmetics, blue conveys textiles. The oval is an omen of death.

Taiwan   There are sex limits on what can be shown in promotional material. Red is considered to be a happy, propitious colour.

Turkey   A green triangle signifies a free sample.

Western countries generally   White signifies purity.

Zambia   People are said to believe that the wearing of red will encour-age thunder and lightning; they do not wear red, and an illustration showing a red clothed figure might have an adverse effect.

United States   Avoid using the national flag in illustrations, and also avoid anything that looks like a swastika. Green is associated with confectionery.

17   Appeal to the senses

17.1   Customer prejudices

In addition to prejudices for and against specific types of package, there are often illogical prejudices which have grown up over the years; these are sometimes well known but may be turned up by research. If such prejudices are found, ways must be discovered to counteract them.

Any design should reflect what customers want, but it may be difficult to ascertain this. The opinions of articulate and intelligent consumers are invaluable, but not nearly so valuable as the opinions of the inarticulate consumer if a way can be found to discover them. Opinions in a poll are not always reliable; a high rating in a poll may not reflect action at point of sale.

Some typical examples of prejudices are:

- Some people are fearful of eating sugar; others consider it a pleasure.
- Some people react to a warning when buying deodorants; others take them less seriously. Different types may react to different designs.
- People tend to dislike commercial messages carried into the kitchen; taking the message off the wrapper increased the sale of paper tissues.
- Many people have pet hates, such as sardines in a tin, or salt that goes hard; avoid pet hates which have a large following.

A prejudice which has become important in recent years is that concerned with the environment; people tend to become prejudiced against forms of package which they think waste natural resources, or which they think may litter the countryside. At one time there was a strong reaction against non-returnable bottles, and the fact that cartons

155

are made from recycled paper has been used as a selling point. These prejudices cannot be ignored but a virtue may be made of them in appropriate cases.

In all cases it is desirable to check that a package is socially acceptable in the light of current prejudices and feelings.

Colour

People do have prejudices against colour, particularly on religious grounds; colours used must be attractive to the majority of buyers. The combination of colours used is particularly important and should not have unpleasant associations.

17.2 Sensory appeal

The appeal of a package may not be wholly visual; vision is only part of the sensory output. A package may create a number of impressions in the mind of a purchaser when it is picked up because of the nature of the package itself. For example:

- A package that is too light may suggest underweight.
- A package that rattles might suggest breakage.
- A package that feels hard and cold may put people off.
- A package that is warm and yielding may be particularly attractive for the right product.

There are other impressions which may be adverse, or the contrary. There are often package style conventions, particularly with cigarettes, and people will shy away from a different type of pack; it may be possible to wean the customers to accept a new type but this should only be attempted after careful thought. On the other hand, people often have deep seated prejudices against forms of packaging and these need investigation. For example:

- Studies of the market for bleach showed that people were afraid that the bottle might leak, the problem was solved by a redesigned bottle with a better grip.
- People objected to rimless cans for condensed milk; it was found that people thought that they felt unpleasant.
- In some cases people expect to be able to see the product and do not take kindly to packages which do not allow them to do so.

There are often other indefinable factors which may increase the appeal of a package. These may be discovered by research but are often discovered by chance. For example:

- The sales of vitamin tablets were improved by a jar which did not look out of place on the breakfast table.
- There is often an affinity between product and package; safety matches and their box are a good example.
- The utility function of the package may have consumer appeal, such as kippers in a boilable bag.
- Tomatoes will sell best in wooden boxes with tissue wrapping; other forms of packaging damage sales. This is, perhaps, because people are conditioned to 'feel' tomatoes.

Colour

Sensory appeal is concerned not with colour but with the feel of the package.

17.3 Product appearance

The appearance of the product within the package is an important factor in sales appeal. Glass bottles and jars, for example, allow the customer to see the product and they make the product look fresher and more attractive. Window boxes, transparent boxes and plastics topped baskets have the same effect, and some products benefit from shrink wrapping in plastics film. This aspect is particularly important when the appearance of the product itself is a factor in sales appeal; many bakery products, for example, sell better when the customer can see what they look like. Other products, however, do not need to be seen and may even be most unattractive when packed; one tea, for example, looks much like another and it is not important for the customer to see it. Coffee, on the other hand, benefits from being seen. In some cases, of course, the product must be protected from light and damp by the package.

The appearance of the product when the package is opened is more important than it might seem at first sight, and a good deal of research has been carried out on this. Researchers have found that purchasers often project themselves into the situation of the packer; one example was a can of shortening where the twirl left by the machine when filling the can gave the impression that it had been filled with extra care.

Some other points that have been noted include:

- How crowded sardines look in a tin can give an impression of quality, or lack of it.
- The hissing sound when opening a can of coffee conveys freshness.
- The hardness of beans felt through a foil sachet conveys an unpleasant feeling.

- A layer of gelatin, aspic or parchment paper may hide an unpleasing appearance.
- An addition may help, for example a sprig of parsley in a bottle of vinegar.
- Packs should be full when opened; the consumer must not feel that he or she has been cheated.

The appearance of the product in the package may often be turned into a selling point, such as 'watch for the deep red colour when you open the tin.' The smell of the product when the package is opened may also be significant; products may not smell as people expect them to do, but in other cases the smell may be particularly attractive.

The first glance on opening a package may well dispose the customer to favour the product, or the reverse, and this may also apply to the appearance of the product *in* the package. With luxury products, poor packaging may detract from the appearance of the product and destroy the luxury image. In the textile industry, for example, clothing or fine lace must not appear creased or 'crazed' in a package such as a see through wrapper. This sort of problem requires special attention to the packaging material.

Colour

Colour selection has little significance as far as product appearance is concerned, except that the colour of the package should not compete with the colour of the product. Where possible, the colour of the package should enhance the appearance of the product.

18 Graphical aspects

18.1 Illustration

This section deals with a number of aspects of graphic design which cut across more than one factor in the design brief and which, for the sake of convenience, have been brought together in one place. The first of these is illustration.

Most packages have some form of illustration of the product, but true to life illustrations do not always convey the right impression and it may be necessary to employ sketches or simulation. Some products do not illustrate well and it may be better to show serving suggestions or an impression of the end result, but it is important that the illustration should not be misleading.

If the illustration includes figures, the age, the type of person and the dress of the person should be carefully thought out. The dress should not date; a woman in a mini skirt, for example, would very quickly date the package. Similarly, a product that is intended to appeal to the young woman should illustrate a young woman and not a middle aged one. The latter would, however, be appropriate when the product claim was based on established usage or on 'what mother did'.

Where food is illustrated in natural colours, take care that the shade is one that people associate with good food. While the sight of a bar of soap in pastel colours is hardly likely to encourage people to wash, the sight of food in natural colours will make people want to eat.

Export packages will usually benefit from illustration, but the illustration must be appropriate to the market; a view of the Thames would mean nothing in Rangoon, but a view of Tower Bridge might. The package ought to show a picture of the product, or a representation of it, because some people cannot read and illustration may avoid the necessity for print. Illustrations which can be understood universally are

helpful, and instructions in an appropriate language can often be supp-
lied on a separate tag or label. Make sure that customers know what the
product is and does.

There are restrictions on illustrations of the human form in some
countries. See Section 16 in this part.

Colour

On most packages products will be illustrated in their natural colours.
Care is necessary with the selection of background colours; they should
enhance, and not clash with, the illustration.

18.2   Identification

The nature of a product may be such that it is offered in a number of
different varieties, flavours, sizes and so on. It may be a good idea to
use different colours for each variation – either the colour of the label or
the colour of the closure. In some cases, graphic design may be different. Some variety may be desirable and one dominant colour may be
a mistake. Some distinction ought to be drawn between small size packs
and larger ones; the latter might be made more attractive to encourage
sales.

The colour used might be associated with the flavour or other attribute, and it may be desirable for the colours to contrast with each other
to prevent confusion.

Different types of paint are often distinguished by the colour of the
can; for example, emulsion paint will have a different coloured can to
gloss paint, or it may have a different coloured band. This idea can be
extended to other products.

The use of different coloured labels to distinguish foods with health
properties is mentioned in Section 15.9 in this part.

Colour can also be used for range identification. If a range of products can be unified by colour and by a design carried across all the packs
in a range, the opportunity to group the lines together in a display is
enhanced and one product helps to sell the others. Range identity requires consistent use of design and logo; if the range can be classified
into product groups, the groups can be colour coded for easier consumer identification.

The design and logo ensure continuity and brand identity; the colour
will add variety to the display, and the colours used should be appropriate to the product or variation. Widely different or even contrasting colours may be used to distinguish between different types of product having the same brand name. The use of closures for identification

is discussed in Section 19.1 in this part.

18.3 Legibility

If a package is to have maximum customer appeal, the legend that appears on it must be legible and easy to read under normal selling conditions. This is a combination of good design, expert typography and appropriate colour combinations and it applies, of course, whether the legend is printed directly on the package or on a separate label or tag. It applies particularly to warnings and instructions for use.

Printing on a package should be simple; the fewest possible type styles will help to avoid confusion but, on the other hand, different typefaces and sizes help to convey impressions, such as delicate, rough, strong and weak. The type must be easily readable close to, as well as from a distance, because many shoppers do not wear their glasses at point of sale.

Consider the following points:

- It should be possible to read the product description under normal display conditions; it should be distinct and related from package to package.
- Consumers generally prefer large lettering for ingredients, especially as many are now health conscious.
- Consumer organisations frown on small print, especially for warnings, cautions and conditions.
- The message should be clear and concise; people have very little time to read what is printed on a package.
- It must be possible to see information when the package is on the shelf; it may be stacked horizontally or vertically.
- Safety warnings and instructions need particular care.

Colour

Consider the following:

- Background colours should be chosen with care; some may make text difficult to read and this could give offence.
- Emotional appeal may have to be sacrificed for greater clarity where warnings and instructions are involved.
- Colours should be selected to convey the appropriate message and then modified to obtain the most readable combination.
- Some combinations are more readable than others; black on white provides maximum readability but combinations should also have emotional appeal and impact as far as possible. Black

on yellow is highly legible but very trying to read and it has little emotional appeal.

- Whatever combinations are used, the reflectance ratio between colours should not fall below 8:1.
- If the aim of text is instructional, colours should be strong and should have good visibility.
- If the aim is readability, colours should convey a pleasing image, should not frighten the customer, and should be visible.
- If the aim is recognition, the combination having maximum visibility is not necessarily the best. Blue on white has good visibility but red on white has better recognition value because red commands the eye to an extent that blue does not.

18.4   Direct print versus labels

At some stage in the design of most packages it is necessary to decide whether to print direct on to the container or whether to use a label. The label has the advantage of being more flexible and can be changed more easily; printed containers can be a headache if there is a product change or a change in marketing policy. The design can be changed more easily when a label is used and, as a general rule, more and better colours can be used.

It is unnecessary to say very much about the different types of labels. Conventional printed labels account for a very large proportion of the market and there are also pressure sensitive labels and heat seal labels, while simple gummed labels are used in some instances. The decision which to use is primarily a technical one, depending on the type of package, costs and similar factors.

The important point to remember is that the label is part of the whole package and must be considered at an early stage in the design process. Most of the comments made in this report about graphic design apply with equal force to direct print and to labels, but the designer must know which is to be used; artwork prepared for a printed container could not be used for printing a label.

Labels have some advantages over direct print and these can be useful from a promotional point of view. For example:

- A label can draw attention to product qualities or features, and these can be changed from time to time, perhaps to coincide with advertising campaigns.
- A label can quote product or company slogans, again as part of comparatively short term promotion.
- A label can suggest different ways in which the product can be

used, or food served, and it is practicable to 'ring the changes'.

- Serving suggestions and recipes can be changed seasonally if required.
- Labels can be related to television commercials or other advertising promotions and provide the link between promotion and point of sale.
- Labels can be coated with inks that change colour in processing, for example to indicate that the product has reached the right temperature.

Colour

Apart from any limitations imposed by direct print, all comments apply equally whether direct print or label is used.

19 Supplementary packaging

19.1 Caps and closures

This section deals with a number of supplementary aspects of packaging which are relevant to more than one heading in the design brief and which have been brought together for the sake of convenience. The first of these is caps and closures.

Every package, apart from a box or carton, needs some form of closure. The type of closure may be dictated by marketing considerations, by the nature of the product, by trade custom, or by availability of closure. The type to be used will obviously affect the design of the whole package.

The most common form of closure is the moulded plastics cap, used for glass bottles, plastics bottles, pots, jars and similar types of package. Typical examples are caps for sweet and confectionery jars, caps for toothpaste tubes and caps for bottles of spirits and wine, although most spirit bottles now have closures rolled from printed foil.

The great advantage of plastics is that a wide range of colours may be used, and this provides a number of interesting possibilities. For example:

- Colour can add interest to what might otherwise be an uninteresting product; it can be decorative in nature.
- A coloured cap for a glass bottle will help it to stand out at point of sale.
- A gold cap conveys an expensive image; other colours can also convey an image.
- A coloured cap can provide more attraction at point of sale.
- If the colours of the label are selected to reflect the colours of the product, a different coloured cap in contrasting colours will provide impulse attraction.

164

- The colour of the closure may be used to identify different flavours, aromas and so on. The closure may be coloured separately from the main bulk of the package.
- The colour of the cap may be changed to reflect changing fashion, whether or not the label is changed. It is much easier to change the closure than to change the entire design. Packages can be updated in this way.
- The colour of the cap may be used to draw attention to poison or other hazards, or to products requiring special care.

In many displays, bottles are stacked in rows and the shopper may be unable to see the label which would otherwise provide interest and variety. Coloured caps not only overcome this difficulty but can also provide primary identification when the label is hidden. For example, appropriately coloured caps may distinguish between red, white and *rosé* wines racked in bottles in supermarkets and wine stores.

Any closure should have attraction, visibility and attention getting qualities. This helps the container to stand out from its fellows; a closure can be identified at a distance more easily than a label.

There should not be any discord between the colours of the closure and the colours of the package, its label or its contents. However, there can be contrast and the closure can have quite a different function to the package in the selling process. If a clear bottle is used, some care is necessary to see that the colour of the closure harmonises with the colour of the contents.

Where a closure is comparatively large, such as a wide mouthed jar, the closure should not overshadow the label and make the whole concept look heavy; this would be less important with screw capped bottles.

If the container has a re-use value in the home, the colour of the closure should be suitable for use in the home.

Many manufacturers produce a series of standard caps in a wide range of colours, and this facilitates the use of colour at comparatively low cost by smaller users of packaging who can thereby have an attractive package without the expense of specially designed units. A typical example would be a retail chemist who can pack his own preparations in standard bottles or jars with a comparatively simple label but with the added attraction of coloured closures. Specially designed caps can, of course, be moulded in any required colour to suit the product or the brand. The bulk of standard caps are used for disinfectants and detergents; some typical examples are

- snap-on caps for food containers: use white or food colours
- plug seal closures for detergents: usually blue or green
- plastics topped corks or screw caps for bottles

- domed caps for bleaches
- caps for collapsible tubes: usually white
- caps for cosmetic pots and bottles: usually pastels.

Colour

Good colours for caps and closures are vermilion, crimson, flame red, primrose yellow, sapphire blue, emerald green; all these have good impulse attraction. White and black are neutral. Pastels suitable for cosmetics include pink, sea blue, lime green and orchid.

Caps may be specifically left neutral to avoid detracting from label design.

19.2 Plastics bottles

For many years there was a prejudice against plastics bottles for packaging but this prejudice has now almost disappeared, especially amongst the younger generation, and research shows that plastics bottles are widely accepted for many products. Generally speaking they are cheaper than glass and lighter in weight, but they are not suitable for all products.

The great advantage of plastics bottles in the present context is that they can be coloured (coloured glass bottles are more limited in range) and the user can take advantage of the inherent attraction of colour. If the bottles are available in a standard range of colours and the user can take advantage of one of the standard colours, costs are very much reduced. Co-operation between moulder and user can produce standard ranges at a very favourable price.

Plastics bottles may also be printed in black and white or in colours. However, if white is used, and printed, all products tend to look alike; self-coloured bottles avoid this difficulty, and colour is often the principal reason for using plastics bottles instead of glass. In practice, about half the output of plastics bottles is produced in colour.

Colour

The principal factor affecting colour choice is usage, and the colour is virtually dictated by the product. The principal uses are:

Cosmetics Colour is an important part of the decoration of cosmetic packaging and is often used for coding purposes, for example to distinguish between aftershave and deodorants. Sophisticated colours are important and should follow fashion trends. Pastel tones are preferred, except for the cheaper types of impulse merchandise.

Pharmaceuticals White, black or neutral for ethicals. Restrained colours for other products, usually clinical in nature.

Household products Colours with strong impulse attraction and compulsion. Actual shades will depend on the product but may have to be suitable for kitchen or bathroom use.

Chemicals Neutral colours are usual, except for household products.

Food Food colours.

Certain products break down when exposed to light, and deeper shades are essential in such cases. Green is particularly useful because it has good resistance to ultra violet light and can be used for translucent bottles; most translucents are in pastel shades, but a translucent dark green is feasible and often used.

Similar remarks apply to blow moulded containers generally.

The most recent development in this field is the use of PET bottles for carbonated and other beverages. These are not so far available in colour, although the moulded base can be produced in colour and caps may be coloured. At present PET bottles are not a particularly attractive form of packaging, but this may alter in due course.

19.3 Wrappings

The term *wrappings* has a number of different meanings in the packaging sense:

- The individual product may have to be wrapped within its container for protection or hygiene reasons. Colour is not likely to be significant here, except in the case of some confectionery.
- The individual package may have to be shrink wrapped or otherwise covered for protection reasons, or to increase consumer appeal.
- Individual packages may be given seasonal wrappings, or wrappings to mark specific events.
- Wrapping paper or bags for purchases are a useful promotional tool.

In the first case, individual products are often wrapped in gold or silver foil to convey luxury, or they may be wrapped in brightly coloured foil or paper to convey a mood of excitement. Other cases require individual study in the light of circumstances.

If a package has to be shrink wrapped or heat sealed, there may be some effect on colour selection. A shrink wrapped package needs stronger colours to show through the wrapping. Heat sealing may impose some restrictions on both functional and graphic design; it may also impose restrictions on the colours that can be used.

The use of seasonal wrappings for Christmas and other occasions is well established. Special wrappings may also be designed to mark specific events, such as a centenary or the opening of a new store. Unlike special packages, which are common for Easter products, seasonal wrappings are usually an addition to an existing package and are removed once the occasion is past. Colours should be appropriate to the season or to the event.

The fourth category is not usually considered to be part of packaging, but wrappings of this nature can add interest and variety and are particularly useful to convey a corporate image. A great deal of free publicity is obtained by wrapping parcels in a distinctive paper or bag which will carry the image wherever the parcel travels. In Japan, wrapping paper is an art form in itself and is often so artistically designed that it is kept as a souvenir.

In a study undertaken for a laundry it was found that the way that shirts and other garments were wrapped, and even the twine used, had a decisive influence on the satisfaction of the customer. Good wrapping conveyed a sense of good service. Even tags and sales slips can be used to convey the same image.

Colour

Colours should be selected according to the image it is desired to convey. Very bright or brash colours are best avoided, but distinctive hues can be recommended and they should be associated with the product or the nature of the organisation concerned.

Part IV
SPECIFIC TYPES OF PRODUCT

1 Cosmetic packaging

1.1 Features of the market

Cosmetic packaging has special features and requirements which justify separate remarks. Packaging is a much more vital part of cosmetic marketing than it is with many other products, and very often it is the package which sells the product, much more than the virtues of the product itself. Most cosmetic packaging will derive from a positive marketing plan, but it will be useful to review the essential features of the market.

In many sections of the cosmetic market increased sales can only be achieved by taking business from the competition. Brand loyalty is often the exception rather than the rule; people will shop around and try something else quite willingly. There is a high degree of impulse purchase in many cases, although some products are often bought after careful sampling. There is an important element of gift purchase. Demand for a product may change overnight, either because of changes in fashion or because of some new promotion.

The importance of packaging and distribution is common to all products and manufacturers are always on the look-out for new and effective forms of packaging. Only those products with effective packaging and a good weight of promotion are likely to find shelf space at point of sale, whether this is the chemist, the department store or the supermarket. Packaging is even more important in the latter case because of the high incidence of impulse purchase.

Cosmetics are essentially a women's product, and even men's toiletries are sold to women. Because purchasers need advice about the use and purchase of cosmetics there is still a substantial sale through retail chemists, although higher proportions are now sold through department stores (usually a shop within a shop), through supermarkets (some of which have their own brands) and by house to house selling. The more expensive products tend to be sold through restricted outlets, mainly to build up 'snob appeal'. Promotion must often be supported by informed selling – hence the growth of the shop within a shop concept. However such selling is often through the better chemists' shops. The customer has to be educated in the use of many cosmetics (particularly treatments) and, because of the expensive nature of the distribution, prices are high. Customers must be given the impression that they are buying good value, and good packaging helps to create this impression. However, it is necessary to strike a balance; expensive packaging can be overdone. Many customers object to packaging which is too expensive, and it may do more harm than good.

With cheaper and more widely distributed products, the principal significance of packaging lies in the fact that all competitive products will be displayed together. Effective use of colour for the package may make all the difference between sales of one product and sales of another, even though efforts are made to differentiate products by means of different shaped packages.

Because of the degree of impulse purchase, displays create a need for colourful and effective packaging. The gift trade is important to nearly all cosmetics – especially the Christmas gift trade – and successful sales depend almost entirely on effective packaging with good colour. Even where products are sold through restricted outlets, colour still has a part to play in ensuring the overall attraction of the package.

With many cosmetic products, colour is the key factor in sales and is the main stimulus to sales, and this applies particularly to lipsticks. Women will tend to try different coloured cosmetics for different times of the day, and at the top end of the market colour choice will be very sophisticated indeed. It follows that if colour is important to the sale of the product, it is also important to the package which is the outward image of the product. This does not necessarily mean that the colour of the package has to match the colour of the contents; this would not be practicable with, say, lipsticks where the purchaser will try out the actual sticks.

However, the colour of the package should create the right psychological image – fashion (as with lipsticks), freshness or cleanliness (as with deodorants) or medication (as with treatments). The colours should be appropriate to the product and to the use to which the purchaser will put the product, and in many cases the colours should be appropriate to the part of the home in which the product will be mainly used, such as bedroom or bathroom. In most cases shades should be soft and feminine.

The selection of colours for cosmetic packaging is an exercise in psychology. The colour of the package, and the outer where appropriate, must not only reflect the colour of the product and convey the right image of that product, but it must also attract attention at point of sale, be different from the competition, and help to spark off an impulse sale. The usage of the product and the perfume of the product are other important factors.

1.2 Notes on specific cosmetics

Lipsticks This market is thought to be near saturation and there is little brand loyalty. Promotion is based almost entirely on new shades, fashion shades and new types of package. It is unlikely that every shade of lipstick will have a separate package, and therefore the colour must set off any shade of lipstick to best advantage and it must also be unobtrusive, although attractive. Effect may be more important than actual hue.

Face powder This is another case where the colour of the package does not usually change with the colour of the product. A powder compact is usually carried on the person and should therefore generally be unobtrusive; it must not clash with clothes. Face powder is a very personal thing; a package colour which complements the human complexion would be appropriate, and it must also convey a soft and gentle image.

Shampoos These are generally sold through grocers and supermarkets as well as chemists and may be contained in glass bottles, plastics containers or sachets. Very often the colour of the liquid itself will be sufficient, but colour for opaque packages should be selected according to the type of product and the nature of the customer; for example, it may be aimed at the older woman, at the young, or at the family. There is a tendency towards a more serious approach to shampoos and very often clinical virtues are claimed; in such cases, the colour of the package should back up the clinical image and associate with the smell. Usage and odour may be more important functions than attracting attention at point of sale.

Perfumes The normal type of packaging is a glass bottle, usually with an outer of some kind, and the colour of the outer will be affected by the nature of the scent and to some extent the name. Although a good proportion of sales of perfume are to men as gifts for women, there is also a strong brand loyalty and comparatively little impulse purchase. The colour of the outer should attract attention to a display but should not be too obtrusive; a brash colour might destroy the image of mystery and exclusivity. The cost of the packaging and distribution of perfume is very high and often forms the major proportion of the selling price; consequently a great deal of attention must be paid to the nature and design of the package.

Skin and other treatments   These will usually be bought by older women and there will be little impulse or gift sale. Women do not like attention to be drawn to their use of treatments and unobtrusive colours seem desirable, although the colours might reflect the nature of the treatment (perhaps a suggestion of medication).

Hand creams and lotions   The greater proportion of sales are through grocers and supermarkets and there is a substantial impulse sale. Colours should reflect either luxury or cleanliness, depending on the usage of the product and the sales theme employed. Certain hand creams are, in reality, barrier creams and intended to be used in the kitchen; kitchen trend colours would seem appropriate.

Eye make-up   Colour of packaging is considered to be less important and will probably reflect the colour of the product.

Hair preparations   Colour in the product is important and should be reflected in the packaging. Certain hair preparations have an impulse sale in supermarkets and the like and attention catching colours are indicated.

Talcum and other powders   Essentially for bathroom and bedroom use and should reflect a soft and gentle image; bedroom colours would be very appropriate. There is little brand loyalty; impulse sales and gift sales are important, but eye catching colours might create too brash an image.

Foundations   Mostly used in the bedroom, and there is a fair degree of brand loyalty. Exclusive but restrained colours are prescribed. Design of packaging is, perhaps, more important than colour.

Bath and other salts   Fair degree of impulse and an important gift sale. The colour of the packaging must attract attention to display and, at the same time, convey an impression of luxury and cleanliness. Elaborate and expensive packages are often employed.

Deodorants   Strong impulse purchase, but too brash colours should be avoided. The colour of the packaging should be related to the sales theme, such as cleanliness or medication.

Toothpaste   Sold extensively through supermarkets and grocery outlets. Although there is a considerable impulse sale, there is also a degree of brand loyalty, and colour combinations should be such that the brand is easily recognised on display. The colour may be related to the sales theme, but toothpaste is often the subject of premium offers, and in this case strong, impulsive colours can be recommended. Yellow green should be avoided for all products put in the mouth.

Sun-tan lotion   Essentially seasonal and impulse. Colours should be related to the outdoors.

Soap   Colour will often be related to the smell, such as pine or flowers. Packaging should reflect the colour of the product, although there is a degree of brand loyalty and colours which can be remembered and recognised are useful. There are certain traditions with soap; carbolic soap is traditionally red or brown, and the latter is also associated with laundry soap.

Men's toiletries   There is a strong and growing market with a substantial gift element. A high proportion of sales are made to women who buy for men, although men are increasingly buying for themselves. Gift packs are useful and should appeal to women, but should also have masculine appeal; no man would thank a woman for a pink pack. Colours should have a clear association with the scent and with the sales theme used for the product, such as masculine, out of doors, luxury or exclusivity. The importance of visual attraction on the shelf should not be overlooked.

Medicinals   Frequently packed in amber coloured containers to avoid the effects of light, but there is no reason why the outer should not follow normal rules.

1.3   The design brief

The specimen design brief outlined in Part III applies equally to cosmetics, but there are a number of points which require special attention. In the following, the numbers in brackets after headings refer to sections in Part III.

Marketing aspects (2.1)

The product   This is obviously the key to the whole cosmetic package concept. Colours of cosmetic packages should generally indicate a quality image; they must enhance the colour of the product; and they should have the right association with the product. A slight change in colour will often make a considerable difference to sales, and in some cases colour may have to suggest changes in the product, for example that it has become weaker or stronger. Colour can also have a negative effect; people may not purchase because they do not like the colour. Pastel colours are generally preferred for cosmetics, but avoid pasty colours. Pink, yellow, aqua, peach, flesh, rose, orchid all have feminine connotations; sea blue, lime green, strong green, red are masculine.

The package   The function of the package is particularly important with cosmetics because in many cases it makes the sale. It may also have a positive function as with lipstick cases; it may have to provide protection, and in many cases it has to convey a luxury image.

The nature of the sale   This depends on the nature of the product but few cosmetics are true impulse sales, although there may be a degree of impulse.

Sales characteristics   Vitally important with all cosmetics; the package may be for the handbag, for the bathroom or for the bedroom.

Selling features   These will have a significant effect on colour selection; the colour of a glamour toilet soap would not be suitable for a soap claiming antiseptic qualities. For most products concerned with skin care gentle colours are prescribed; pale hues are generally associated with gentleness. Blue would be very suitable for a product with antiseptic qualities, but it may be a mistake to overdo the antiseptic aspect. If the product is intended to be a fashion item, use fashion colours. A blue/green/silver combination helps to create a clinical look.

Scent is a selling feature of almost all cosmetic products and the colours used for the package should be associated, as far as possible, with the perfume of the product, such as pine or flowers. Consider:

- Pale colours are generally associated with flower scents, but there are many specific associations.
- Where the product is sold in a number of different scents, it may be desirable to have a different coloured package for each variation.
- If a new perfume is introduced, or an alternative to something already on the market, different colours for the package would emphasise this fact.
- The perfume may be designed for the male but the colour of the package should appeal to both sexes; women are the principal purchasers of products intended for men.
- Consider the association between brand and scent.
- A pine soap would benefit from a pale green wrapper.

Repurchase   Refills may be important.

Price   May have some effect on colour selection; the more expensive the product, the more sophisticated the colour.

Product associations   Follow these where appropriate for both type of package and colour. For example, a detergent type bottle would not be appropriate for a skin product. Certain colours have been found to be particularly suitable for specific products, for example blue for aftershave, white for sanitary products. It may be desirable to question the use of colour; pharmaceuticals may be better without colour. Consider:

- There are traditions; red is traditional for carbolic soap.
- A product which is concerned with cleanliness should have a package in clean, cool colours.
- People do not like to put yellow green in their mouths; it is unsuitable for toothpaste and may be for other products.
- Cotton tipped swabs in white and pink sold well but the same product in chartreuse did not because it reminded people of dirty nappies.

Product image (2.2)

Product identity is particularly important with the more expensive cosmetics; the package must be 'different', and a distinctive shape of package may be a useful identification. It may be desirable to avoid an 'English' look unless the product concerned is a recognisably English one; there is snob appeal in foreign associations.

Soft and gentle colours are usually desirable; harsh colours may convey the wrong image, particularly with skin care products. A supermarket type of package is not suitable for high grade toiletries, but colour can convey a Rolls-Royce image for products sold at Ford prices.

There may be a house colour, brand name or trade mark which provides a unifying factor throughout a range, but this need not prevent the use of colour for identification.

Physical characteristics (2.3)

Physical form This decides the nature of the package and whether a bottle, sachet or other container is required. A carton may also be necessary for protection or promotion.

Physical characteristics These will affect the nature of the package. It must not leak.

Wrapping A tinted film wrapper may be necessary to prevent product fading.

Emptying A dispenser or other suitable device may be required.

Evaporation This is particularly important with liquid products such as scent.

Protection   Expensive products need adequate protection against breakage.

Toxicity   This may be significant with some products.

Product appearance   A basic question is whether the product needs to be visible; if so, the colour of the package should enhance the colour of the product.

Types of packaging (2.4)

This will depend on the nature of the product, and there are often established conventions about types of packages for cosmetics. For example, older people are not so keen on plastics bottles, although teenagers approve of them. Special effects may be useful in some cases, such as pearlescent, tortoiseshell, gold flash, marble effect, translucence.

The customer (4)

Age   Teenagers prefer simple, clean designs and like striped patterns and abstracts. They also like large areas of white and bright colours; pastels are second choice. Older women tend to be influenced by Paris, younger women by New York, and this may have an influence on colour selection.

Sex   Most cosmetics are bought by women and benefit from feminine colours, but men's toiletries, although often bought by women, require masculine colours. A feminine colour, such as pink, would be most inappropriate for men's toiletries.

Purchasing habits (5)

Presentation   Purchasers may expect to be able to see the product; a transparent package may be indicated, but the colour of the label should enhance the appearance of the product.

Usage of the product (6)

Where used   This is particularly important with cosmetics. For example, if the product is to be used in

- the bedroom, soft, gentle colours are indicated
- the bathroom, clean, cool colours are indicated
- the kitchen (such as hand lotions), a clinical look is indicated
- the workshop (such as barrier creams), strong, masculine colours are probably best.

If the product is to be carried in the handbag, the colour of the package should be unobtrusive and not clash with women's fashions.

Suggestions for use   It should be quite clear what the product is and does. Instructions for usage may be important.

Selling conditions (7)

Type of outlet   If the product is sold mainly through chemists the colours of the package should normally be more restrained. Brash colours suitable for the supermarket are not required. Products sold through restricted high class outlets normally require sophisticated package colours; colour choice should be subtle and aimed directly at the market in question.

Method of promotion (11)

Gift sales   The gift trade is very important to certain types of cosmetics, especially men's toiletries. It is desirable to analyse the nature of the trade for a specific product and to identify purchasing habits before deciding which colours will be most suitable. Special colours for gift packages are often useful, and are normal for Christmas packages.

Sales theme   The colours of the package should support the promotional theme, such as medication or cleanliness. Colour and personality might well be used as a sales theme for soap. Cosmetics are often sold on snob appeal, and colours selected for packages should support this; the fashion aspect is particularly important and a sophisticated product should always follow the dictates of fashion.

Technical features (13)

Type of closure   Gold tops or caps dominate a display under normal lighting and are useful in appropriate cases; they help to upgrade an image.

Labelling (15)

Identification   If the product is produced in more than one colour (such as lipsticks), flavour, scent and so on, it may be that different colours are desirable for each variation although a neutral hue may be chosen in all cases. A house colour, brand name or trade mark may provide a unifying factor throughout a range, but this need not prevent colour being used for identification purposes.

   If the product is part of a range, for example hair care preparations, group the colours of the package according to function and have variations of colour within each group. It is useful to use similar colours when products are intended to be used with each other, such as toning lotion with shampoo, but this does not mean that the colours have to be exactly the same.

Legibility (18.3)

An up to date typeface is usually desirable unless the product has an old fashioned image.

Product appearance (17.3)

Consider how the product will look when the package is opened; the colour of the lining of the carton may be particularly important with face powder, for example. Special care is necessary with the association between product colour and package colour when the package is transparent or translucent.

2 Confectionery packaging

2.1 Features of the market

In the present context confectionery means sugar confectionery, chocolates and so on, but not flour confectionery such as iced cakes or biscuits, which are included under the heading of bakery products in Section 5 of this part.

Apart from boxed chocolates, most sugar confectionery is sold in comparatively small packages, which are racked or otherwise grouped together on the retail counter. Most sales are impulse sales. To a large extent purchase will be guided by what the customer sees and the extent to which they are attracted by the individual package, such as a wrapper or a pouch. The package creates the sale to a much greater extent than with food, and good design, good colour and good combinations of colour are a vital necessity if the package is to have any chance of being picked out of the welter of conflicting lines.

The importance of well thought out packaging is most marked with count lines such as candy bars and the like. At one time confectionery was packaged in boxes which were kept under the counter and only brought out as required; display consisted of a few samples on a saucer in the window. Later, the glass bottle provided a means of packaging which protected the contents and looked reasonably attractive, the attraction being provided by the product itself. Glass jars are still used, but the limited amount of shelf space in the average shop means that products packed in a bulky jar may never catch the customer's eye and, in addition, retailers are increasingly disenchanted with the time taken to weigh out small quantities. The trade has moved in the direction of self-selection; the package has become the major sales aid, and the individual package appeals to both retailers and customers because it is much less trouble to buy and handle.

Before any package can create a sale it has to be seen by the customer and therefore has to secure shelf space. This is a matter of discounts, representation and skilled merchandising, including display aids of various kinds. Display outers may be used, but as lines proliferate the appearance of the individual package becomes even more important. The individual package is particularly important to boxed chocolates, where the customer buys the box rather than its contents. Boxed chocolates could hardly be sold in plain wrappings, and in this case the package is more important than the product. Although some customers resent expensive packaging, few are likely to do so when buying chocolates.

The confectionery package is an integral part of that section of the selling process which is described as the struggle for display space, but it also has to force its attention on the consumer and create a sale. Some confectionery lines are, of course, advertised extensively but in the main sales depend on customer attraction at point of sale. The package has to create attention of itself or by recognition, and then prove sufficiently attractive to impel the customer to pick it up and buy it. When the product is advertised extensively, the package is even more important because it creates the image that links advertising and point of sale.

All this is not very different from the function of the package in any other market sector, but there is a greater emphasis on the package in confectionery selling because of the importance of impulse sales.

2.2   The design brief

The specimen design brief in Part III applies equally to confectionery packaging and points that require special attention include the following. The numbers in brackets after headings refer to sections in Part III.

Marketing aspects (2.1)

The product   The nature of the product will have a considerable influence on the nature of the package. A plastics bag is suitable for peanuts but not for chocolates; a picture box will sell chocolates but not jelly babies. The colour of the package must reflect and enhance the colour of the product inside. The colour should also be acceptable for things that are put in the mouth, and attention is drawn to colours suitable for food packaging.

The package   In many cases, as with a box of chocolates, the package makes the sale, but with confectionery items the primary function of the package is to attract attention at point of sale.

The nature of the sale   The great bulk of confectionery is an impulse sale, hence the importance of attractive packaging.

Selling features   These can include taste, flavour or quality, all of which can be emphasised by suitable colours. Colour can also convey appropriate associations; for example, an Easter package should include plenty of yellow and green, the colours of spring. Value for money is often a selling feature, and in such cases it would be desirable to use colours that make the package look as large as possible.

Product associations   Certain colours are traditionally associated with specific types of confectionery and therefore help to create an image; red and white striped peppermint sticks are a case in point. Generally speaking, the colour of the package should be related to the taste or flavour of the product or its ingredients, for example brown for nuts or green for peppermint flavour. Pastels suggest sugar confectionery, and the so-called 'candy colours' are in fact a selection of pastels.

Product image (2.2)

Product brand image   Colour is often used in the confectionery field to create a product image. 'Cadbury purple' is a practical example which has become a brand image, despite the fact that purple is not usually considered a good colour to go with things to eat. Another similar example is Rowntree's use of black for their Black Magic. Where the package is designed to promote a brand image, it must remain relatively unchanged for long intervals. A novelty pack, on the other hand, has a short life and can use transient colours.

The market (3)

The nature of the market   This needs careful analysis. Much of the packaging of cheap count lines is neurotic and screams for attention, but this may not appeal to the type of market which is the best outlet for the product. While a neurotic design catches the eye, it does not always follow through to a sale; it may stand out quite well at point of sale, but the package tends to 'cut itself into small pieces' and lacks unified appeal. More restrained colours would add more grace and persuade more people to pick up the package.

   Restrained and dignified colours are recommended for a high class market. An elaborate package is a 'must' in some markets but will lose sales in others.

Speciality markets   Confectionery products often appeal to, or can be made to appeal to, a specialised market, and this requires consideration in both package design and selection of colour.

The customer (4)

Age   A confectionery package must appeal to the age group at which the product is aimed; many lines are aimed at children or teenagers who will react best to bright and up to date colours.

Purchasing influence   Who is it? An Easter package is a fairly simple conception, but who buys it: the child, the parent or the auntie? Parents will buy what they think will appeal to the child, but children have their own ideas and are attracted by colours which do not appeal to adults.

Purchasing habits (5)

The reason for buying and the normal purchasing motivation is significant in the confectionery field. Much confectionery is bought as a gift; there may have to be one package that appeals to a husband taking a present home to his wife, and another for a man giving a present to his girl friend. It may be necessary to decide which is the more important market.

A package may be bought because it is suitable for the handbag or pocket, and more restrained colours might be desirable in such cases.

Selling conditions (7)

Attention must be paid not only to the immediate wrapping but also to the carton, outer and display case; all should eliminate the need for other display aids at point of sale.

Method of promotion (11)

Special offers   Many confectionery products lend themselves to special packs for special occasions. Christmas is the most obvious example, and colours should be appropriate to the occasion.

Sales theme   If the theme is 'value for money', use colours that make the package look as large as possible. Milk chocolate would benefit from colours that suggest milk.

3 Hardware packaging

3.1 Features of the market

Hardware is a difficult sector to define because it can include such a wide variety of products and many different types of packaging. However, for present purposes it divides itself into small items such as screws, hooks, plugs, small tools and the like, and into larger items such as housewares, kitchen utensils and even certain types of space heaters. It also includes paint and garden products.

The hardware trade has changed radically in the last few years, and impulse purchase has become a vital part of sales. In the past, hardware manufacturers were slow to realise the importance of packaging and retailers were afraid of it, but now all sections of the trade recognise the value of a good package, especially in those areas where there is little to choose between competing products. The screws of one manufacturer are very like the screws of another, and the package may be the only competing feature. The colour of the package adds attraction to what is, in many cases, a rather colourless product.

The principal function of hardware packaging is to sell the product, but it must also protect the product and assist in the process of distribution. There have been many complaints over the years about the inadequacy of hardware packaging, and this is mainly due to lack of thought at the design stage. Some typical complaints include:

- paint tins which do not identify the colour of the contents
- packages of light bulbs which do not indicate whether the bulb is pearl or clear
- packets of screws which do not say whether they are brass or chrome.

Most of these complaints boil down to the fact that the contents of the

185

package cannot be identified when it is on the shelf. For example, colour identification on the lid of a paint can is useless when the cans are stacked. Most hardware lines are subject to increasing competition from overseas sources and the package must compete with the best of European design, both graphically and functionally.

3.2 Hardware display

The most significant feature of present day methods of hardware retailing is the development of what may be described as prepackaging, necessitated by the growth of self-service selling and the practice of selling hardware items in non-hardware stores. The more usual forms are:

- In bags with headers. Normally for loose items, and with the disadvantage that the headers tend to wilt.
- In blister packs with card backing. Very suitable for single items and generally preferred by retailers.
- In shrink wraps. These have disadvantages.
- On cards, either singly or in multiples. The multiple form is considered to be outdated because the cards look unsightly as the items are removed. Single cards with facilities for hanging are preferred.

Where cards are used they are frequently contained in outers of a dozen, and this has the advantage that the retailer does not have to hold large stocks. On the other hand, the retailer does have to find wall space or floor space for the displays.

Prepackaging of this nature generally sells on impulse and the main selling feature is convenience, but the method is often adopted without really considering whether the product is suitable for it. Consider the following points:

- The package must be strong enough for the product.
- It must be attractive and create sales.
- The size and shape must be as uniform as possible, thus making for a more attractive display.
- There must be facilities for hanging, or some other method of display.
- There should be a brief description of the product and notes on its use.

The method of packaging described above applies mainly to smaller items such as nails and plugs, but the display aspect of the packaging of larger items is just as important in attracting attention at point of sale.

Cutlery is a typical area where there is a much greater degree of buying on impulse than there used to be, and stainless steel cutlery will benefit from an attractive box which displays it to best advantage. The box permits colour to be introduced into the display of an item which is itself colourless, and the box can also help to establish a brand image.

One of the items taking up the largest amount of space in the average hardware store is paint; well designed paint cans can contribute to sales, although paint is not normally considered to be an impulse purchase. The design should convey a dynamic and pleasing image and the colours of the can should reflect the trends of the day, because these are the images of paint itself. The design should also attempt to convey an image of many colours, although reproduction costs may militate against the use of too many colours in the design; coloured dots might be used, for example, to convey a multiplicity of colours. The colours of the contents of a paint can should be easily identified at point of sale, especially when stacked. One paint manufacturer uses three bars of colour around the top of the can, different colours being used to identify the type of paint (gloss, emulsion and so on); the colour of the contents appears on a band around the bottom of the can. This type of marking is visible under any conditions.

3.3   The design brief

The points that require special attention when designing hardware packaging are listed below. The numbers in brackets after headings refer to sections in Part III.

Marketing aspects (2.1)

The product   There are few colours that are particularly associated with hardware, except that blue is associated with steel and green is associated with garden products and with products having a cleaning function. Many hardware lines are themselves coloured, and items like kitchen knives will have coloured handles; the colour of the packaging should enhance these hues, which are often an important selling point. The colours of packages containing small appliances and other kitchen products should generally follow kitchen colour trends.

Selling features   Some hardware products have a distinctive smell which may be a selling feature; the package colours should associate with the smell.

Illustration   This may be particularly significant with hardware items; it is common practice is illustrate usage by a picture or to illustrate the end product as with, say, garden seeds. In the latter case, great care should be taken not to make or suggest claims which cannot be justified in practice. Illustration may present problems with some types of packaging and it may be necessary to enclose an instruction sheet or leaflet.

Physical characteristics (2.3)

Physical form   The weight, size and shape of the product needs to be considered, and how it is to be loaded into the container. Some method of carrying such as hand grips may be required, at least for larger items.

Protection   Present day methods of hardware retailing require that most items require protection while waiting for sale. Many lines, such as heating appliances, have a decorative function, and this needs to be protected up to the point where the appliance is installed at the ultimate point of usage; there may well be vulnerable points that require special protection. Fitments may be required inside cases; irregular shaped objects require supporting fitments. Saucepans wrapped in brown paper will not stack. Where articles have multiple components, the components may scratch each other unless each has adequate protection.

Types of packaging (2.4)

The hardware sector is notable for the wide variety of types of package which have a place in it. Some examples:

- skin packs
- vacuum formed packs
- cardboard boxes
- metal boxes
- plastics bottles
- loose in display case
- clear packs
- wrappers
- fiberite cartons
- cans
- glass bottles
- bubble packs
- clear wrappers
- plastics wallets
- plastics jars
- glass jars.

Most of the packages are designed for display purposes, although some are purely protective in nature. Some are primarily intended to form a display background to the product, such as cutlery cases.

Usage of the product (6)

A container may have a use in the home as well as a display or presentation function; it would be desirable to consider the part of the home in which it might be used and to select colours that are appropriate. It will also be useful to consider how the customer might use the container, for example for storage or dispensing.

Selling conditions (7)

Shelf display Display characteristics are the most important factor affecting selection of colour for hardware packaging; in most cases strong impulse colours are desirable so that the package will stand out at point of sale. When packages are stacked, each package should be separated from the other in visual terms. A stack of cartons of household foil, for example, should not become a blur; each carton should be clear cut and have its own identity.

Distribution conditions (8)

Channels of distribution It is very useful to investigate the way that the package is handled from the time that it leaves the factory until it reaches the home of the ultimate purchaser. Loads may be mixed, for example of different sizes, and they may have to be broken or subdivided at some stage of the distribution process. The package may have to be opened and reclosed at some stage.

Storage conditions There are many complaints about hardware packaging not protecting the contents properly during handling and storage. This applies particularly to larger items such as appliances which are very liable to damage and may be difficult to stack.

Information Most cases containing hardware need to be marked in more than one place; colour coding may assist identification in store. More complex items need to be adequately labelled so that all concerned know exactly what is in the package. Identification of the contents must be easily visible when the packages are stacked.

The retailer (9)

Needs of the retailer Storage, display, product access and carry home functions are all important to the retailer. The retailer may have to break bulk in some instances and the packaging should allow for this. Heavy items require a handled package.

Display All hardware should be packed to facilitate display, and it is very necessary to consider how the package will be used at point of sale. A frequent complaint from retailers is that identification cannot be seen when containers are displayed; this applies particularly to colour identification on paint cans.

Method of promotion (11)

Special sales themes Promotion of kitchen utensils and housewares may well be based on fashion in the kitchen; the colour of the product will be selected for this reason and the package colours should mirror it. The colour of the package may have to be changed when the product colour is changed to reflect current trends. Where the sales theme is concerned with other parts of the home, the colour of the package should reflect appropriate trends in the same way.

Labelling (15)

Identification   This is important with many hardware products, for example to distinguish the colours of the contents of the can or package and to indicate size or other attributes. Special care may be necessary where a case contains a selection of articles in assorted colours, an arrangement which is common with plastics housewares.

Colour   If colour is used for coding or identification purposes, the code should be easily understood and there should be good contrast between the various colours used.

Instructions for use   In almost all cases hardware packaging must bear a description of the product, instructions about usage, and possibly suggestions for using the product to best advantage. It may not always be possible to provide the information on the package or on the product itself, and a separate label or tag may be required. Tags are frequently used for larger items such as appliances.

Package contents   There are many complaints about labelling which does not identify the contents of a package. Adequate labelling should include

- the product name
- the size or grade of the article
- the material from which the article is made
- the finish
- the colour, where applicable
- the quantity in each package
- in appropriate cases, whether all components are included.

4 Textile packaging

The packaging of textiles such as sheets, table cloths, napkins and the like, and also articles of clothing such as shirts, has become very much more important in recent years, mainly because of changes in retail trade which have put a premium on presentation. The use of distinctive packaging helps to establish a brand image and enables the manufacturer to use colour as an effective sales medium. Packaging also helps to avoid soiling and crushing in the distribution process.

The appearance of the product in the package, or immediately on opening the package, is particularly important. It is vital to avoid a crushed or creased look when the product is in the package, especially if a 'see through' package is used. Many textile products are wrapped in plastics film, and if this is tough enough to withstand printing it tends to become brittle by the time that it reaches the shelf. Soft films tend to split during handling and cannot support printing; therefore tags and swing tickets have to be used for washing instructions and the like. Special films are available.

Packaging must enable the customer to identify the colour of the contents; where film is not used, a window or clear topped package is appropriate. The colour of the package itself should identify the manufacturer. Fashion is important with all textiles and up to date colours are essential. The colour used for the package may, however, reflect a sales theme, for example fashion, restful colours or country colours.

Hosiery, such as stockings and tights, is essentially a fashion item. The colours used for the packaging should be fashionable too but they must not smother the colour of the product; they should set off the product colour to best advantage. It is particularly important that the colour of the package does not mislead the customer about the colour of the product; there is always a risk of after-image. The nature of the market is significant; sophisticated colours are required for high class markets. The sales of the product may depend on a luxury and expensive image.

Much lingerie is bought by men for women and black is always a safe colour to use. A man will almost always be attracted by black lingerie, although a woman will seldom buy it for herself.

Yellow can be recommended for packaging of children's wear but avoid green. Pink and blue are also suitable.

5 Food packaging

5.1 Features of the market

In the present context, food means meat, vegetables, fruit, groceries and bakery products. Confectionery, that is sugar confectionery, is discussed in section 2 and beverages other than tea or coffee in section 8 of this part. Food packaging is particularly important because of the conditions under which food products are sold; present day self-service conditions require that the food package should make the sale under intensely competitive circumstances.

Food packages are not bought for themselves but for what they contain. In this they differ from, say, chocolates where it is the package that is bought rather than the product. The food package needs a certain compulsion to ensure that one brand is bought rather than another.

There have been substantial changes in the nature of the food market in the 1980s, including growth in the demand for health foods, foreign foods and convenience foods. All these have some effect on packaging. Snack foods are increasing and the instant hot meal market is growing, and these require new types of packaging. Three major growth trends have been identified in the 1980s:

- concern with health, wholesomeness and diet
- interest in foreign, delicatessen, exotic and ethnic foods
- desire for convenience in preparation and presentation.

Health foods

This is a rapidly increasing market which includes both health foods proper and health orientated products, such as those concerned with slimming. These foods appeal to the AB market and there are many specialist stores; health foods also sell through chemists' shops. Both outlets require some rethinking about packaging. The main categories in the health food market are

- whole foods including cereal products
- vitamin supplements including slimming foods
- herbal remedies and herbs generally
- natural cosmetics which may be sold alongside food
- homeopathic treatments.

There is a trend towards light foods, more seductive, more attractive and appealing to dieters and non-dieters alike. Libby emphasise reduced sugar; Del Monte emphasise nutrition value; Tesco are labelling all their own brands with a nutrition symbol. Most light foods are low calorie, although there is no sale for low calorie desserts. Some manufacturers are concentrating on white for packaging to emphasise the health aspect, but in general the aim is to make foods look appetising and healthful rather than medically dietetic. For this reason there is likely to be a good deal of package redesign.

Foreign foods

The UK is a healthy and growing market for foreign foods including French, Italian and German. In the last case food accounts for 11 per cent of total exports and they concentrate on premium priced speciality products that are unique to Germany. The shift away from manufactured products towards natural foods should give the Europeans their best chance.

The chief significance of this from a packaging viewpoint is that food producers have to react to the foreign success in changing British tastes and persuade the consumer that British food has much to offer. They have to react to changes in life style and eating habits, and this means new types of packaging and improved design. The British consumer is becoming choosy and more discriminating, and is demanding variety and higher quality; there is growing dissatisfaction with the standard product in a number of areas. The change in eating habits towards more natural, less processed foods should give traditional British foods a new opportunity, and this must be reflected in packaging.

Convenience foods

This is a wide term which includes frozen foods, continuing to expand at the expense of canned foods; speciality foods, such as pizzas and prepared salads; and the so-called convenience desserts, which appeal primarily to children. All these require new and improved forms of packaging and, in many cases, a rethink about design. Consumers need education in the use of many foods and this is primarily a matter of more informative labelling, supported by advertising.

Fresh foods

There is a growing trend to fresh foods, and these require packing. Products showing interest are prepacked salads, pâtés and pizzas, but the main growth areas are short lived dairy products, delicatessens and pastry goods.

Colour

Notes on colour in relation to food packaging will be found in Section 6 of this part.

5.2 Specific foods

The following are some notes on the packaging of specific foodstuffs.

Baby foods

Although baby foods are also sold through chemists' shops, they have been treated as food products in this guide. The package should be aimed at a universal market because there is no class significance in purchase. However, it has been found that there is a difference in the purchasing habits of white and black mothers; the latter tend to buy the more expensive brands, have much greater brand loyalty and tend to rely more on package recognition. The name of the brand may not register at all but the package does. The white mother relies more on the brand name, and if she is the mother of a happy baby she will not change the brand.

The package should look neat, discreet and hygienic, and should be secure. It should convey a high class, quality impression and be up to date and modern, but not too modern, because mothers do not like untried things. Nor should the packaging be overdone, because customers may think that money is being wasted. A good, recognisable, visual image is recommended, and shape and form should be rounded. A baby is, after all, round.

Many baby foods are bought on recommendation and visual continuity is most desirable because of repeat purchases. The mother must be able to recognise the package that she bought last time; continuity of shape and colour are also desirable. What the mother uses in hospital she will also use at home, and recognition is vital.

Robinson say that the package design is as important as the product itself; there must be a strong shelf image with a clear differentiation between each product.

In the marketing of breast feed substitutes, regulations require that the superiority of breast milk must be emphasised at all times.

Biscuits

Biscuits are largely bought on impulse and the packaging can create
sales, although it can also depress them. Biscuit packaging is frequently
criticised for being too difficult to open or for failing to protect the contents. Broken biscuits are a perennial source of complaint to consumer
organisations. The package ought to be attractive, keep the contents
crisp and fresh, and should be easy to open without causing wastage.

Canned foods

Only pet foods are showing any growth in this area, although canned
pasta is said to be doing well. Smedley repackaged their range of
canned products to show and feature an English theme and to include
colour pictures of the product; they also expanded their range of
canned salads to take advantage of the health food trend.

Dairy products

Because of declining quotas for butter and cheese, New Zealand laun-
ched real cream in an aerosol container – the first use of aerosols for
food products. Cream in long life cartons downgrades the product.

Desserts

Convenience desserts are showing a decline because of changing adult
life styles; less formal eating habits mean that people are dropping a
sweet course and eating fresh fruit instead. Sweet foods have a wealth
of health, fashion and social factors stacked against them. Children are
the mainstay of the dessert market and packaging should possibly be
directed at them, particularly as the frozen and chilled sector is the most
successful; ice cream, yoghurt and frozen gateaux have all shown increases, and this puts a premium on attraction at point of sale. Sales of
jelly, canned desserts and whips are all stagnant.

Frozen foods

The market is buoyant and consumers seem willing to pay for quality,
convenience and ease of preparation. The most successful products are
the more complicated ones which are difficult to make at home. New
product development has been one of the factors in market growth and
producers are looking for added value products; these require more elaborate packaging.

   Another factor in market growth has been home freezers. Two-thirds
of households now have them; the trend is to upright freezers rather
than chest type, and to combined fridge/freezers. This has packaging

implications. There is a greater use of prepared frozen foods rather than freezing of home prepared foods.

The major food chains have allocated more space to value added foods and consumers are likely to buy frozen foods on impulse if they see them on display; new product ideas have stimulated consumer awareness, and this puts a premium on packaging that catches the eye. Other points:

- There is a trend away from frying to the greater use of grill and oven.
- There is demand for low calorie speciality meals in frozen form.
- Peas account for about half the vegetable market, which is itself about half of all frozen foods.
- Indian style frozen foods have been introduced by Sharwood.
- Fish fillets have taken over from fish fingers.
- Frozen cakes and desserts are thriving.

There is intense price competition in all sectors, and the growth of own label brands and independents has sliced the market share of brand leaders. The latter have a major problem in ensuring identification in the cabinet.

Pastries

The package should be more than a convenient way of taking the product home. The average carton does very little to create sales, but there is no reason why it should not contribute more and enhance the appearance of the product, especially where the product is sold through supermarkets.

Tea

Packaging of tea is the critical element in the marketing mix. Twining are sacrificing their logo in the belief that by doing so they can create a new market in speciality teas. Tea consumption overall is falling but speciality teas are growing. Twining's existing package did not look as stylish as the competition; Horniman's package conveyed quality, tradition and up to date appeal. Twining's new pack was designed in black and gold (considered stylish) with a delicately drawn cameo in colour for each type.

Vegetables

Prepackaging has advantages in some sectors, but research into suitable outlets is necessary and contents of prepacks may have to vary according to the market and the location. Branding is growing and is needed to counter the superiority of foreign branded products. British vegetables tend to fall down on quality, consistency and presentation and there is no organised system to cater for large buyers. Sales through Covent Garden are decreasing as supermarkets buy direct; Asda say that their customers recognise vegetables as Asda quality.

Prepackaging could be made more acceptable to the public if more thought was given to the package and to the appearance of the package. There is still some prejudice against prepacked produce, and this stems from a number of factors including the traditional way of buying produce by sight and feel. Customers tend to think they are being 'done' when they see familiar products in 'fancy dress'. This feeling is particularly marked among older customers. Some customers also object to buying larger quantities than they really need.

There is no reason why people should not accept packaged produce if confidence can be established in a brand name. This requires an attractive and easily recognisable package containing a carefully selected product of good quality. A brand symbol would be particularly important in establishing acceptance of the brand because the shopper will tend to look for apples (or whatever) first and for the brand symbol second. Anything else is less important.

With some types of product a standard container is used, and in such a case symbol and design are vital because they are the only distinguishing features. There is little point in discussing the relative merits of various types of container or package, but a rigid package lends itself to decoration and design. Obviously there must be consideration of the method of packaging, handling during distribution, protection of the product, and the needs and convenience of consumers. The last includes size of package.

It is often said that the consumer must be able to see prepackaged produce, that is the package must be transparent or have a window, but this might be less important if customers could be persuaded to depend on the brand image. Until this is achieved, the appearance of the product within the package is vital.

The shape of the package will usually be determined by mechanical and physical factors, but it should be mechanically sound and as attractive as possible; anything that looks 'sharp' or difficult to pick up should be avoided.

Potatoes require special consideration. If wrapped in film they require protection from strong light to prevent their going green, and they also require to be kept at constant temperature. A kraft paper outer will provide the necessary protection, and they should be kept in this until sold; the outer provides scope for brand labelling. Carrots, onions and other root crops are not so temperamental.

5.3   The design brief

The design brief outlined in Part III applies to food packaging just as it does to any other type of packaging, but there are a number of points which require special consideration in the food field. In the following, numbers in brackets after headings refer to sections in Part III.

Marketing aspects (2.1)

The product   The nature of the product is, of course, the key factor because this governs the type of packaging, the nature of the design and every other facet. In this respect, food is no different to any other product.

Illustration   Pictures of food are popular and recipes often get used. Illustrations showing the finished product are best in most cases and they should titillate the appetite but, where appropriate, it should be made clear that the illustration is a serving suggestion. Food shown in association with a good table setting is considered to be an asset.

Product ranges   Many food products are sold in a variety of flavours, mixes and so on. If such a range can be unified by a design carried across all the packages, there is an opportunity to group all the lines together in a display; one product helps to sell the others.

Physical characteristics (2.3)

Food manufacturers have to take account of a number of characteristics when designing packages, including perishability, freezing, canning and distribution regulations. If transparent containers are used, the product may have to be protected from exposure to ultra violet light. Hygroscopic substances such as soup mixes can often only be protected by aluminium foil; this also has the advantage that it reflects heat and helps to keep the product cool.

The customer (4)

In general the nature of the customer is not significant, but the health food market is essentially an AB market and good class packaging is prescribed.

Purchasing habits (5)

Certain products are regularly bought in small quantities; others are bought at irregular intervals and stored for long periods. The common practice will affect the nature and size of the package.

Distribution conditions (8)

Location of price spots is important; the best place is on the end of the package. Campbell use a one layer case for soups which exposes all the tins for price marking, and the case is also easier to handle than a larger one.

Method of promotion (11)

The health food angle is now particularly important and may be reflected in the design of the package by showing calorie count, fat content and so on. Other themes which may be used, and which should be reflected in the packaging, include freshness, convenience and culinary expertise. If the selling theme of the product is convenience – 'take a packet home for your supper' – no one wants a paper bag.

Technical features (13)

Easy opening devices and reclosures may add attraction to a package and be a useful sales point, especially for those foods that may not be used immediately.

Product appearance (17.3)

The appearance of the product in the package and its appearance after the package is opened is particularly important with food products. Window packs, plastics topped baskets, glass jars and bottles have the virtue of allowing the customer to see the food, and they help to make it look fresh and attractive.

Consumers often complain about containers not being properly filled and, if there is a reason, this may have to be explained on the package. The customer may also have to be prepared for, or warned about, what to expect on opening the package, such as 'smell the fresh odour of green peas.' Smell is very important and may have to be controlled or explained.

When a can opener cuts into the contents of a can there is often an unfavourable impression, and the housewife may feel that the contents of the can are not clean. Anything that can be done to improve the appearance of the food when the container is opened will be helpful and, if possible, appearance should be made an additional selling point. Some food products are particularly unappetising at first sight, such as canned herrings, and an aspic layer, or even a layer of parchment paper, will improve the first appearance. Another suggestion of this nature is to place a carefully selected sample biscuit at the top of a box of biscuits so that the impulse to taste can be gratified.

6   Colour for food packaging

6.1   Introduction

The importance of the relationship between colour and food, and the
high proportion of the total packaging market represented by food
packaging, justifies discussion of the subject of colour for food packaging at rather greater length than colour for other specific types of packaging. This is because colour is always part of food and, however good a
food may be, people will not buy it if they do not like its appearance.
When people cannot see the actual product, they will make a judgement based on the package and the appearance of the package, and it
follows that the colour of the package may be just as important to sales
as the colour of the food itself. Food packages in general are not purchased for themselves but for what they contain (although there are ex-
ceptions to this rule).

There are few colours that are really suitable for food packaging, because there are a limited number of colours that are psychologically
pleasing in relation to food. Colour is always part of food; it is a highly
visual element to which human emotions, minds and palates are extremely sensitive. A good meal is pleasing to the eye, and the colours
selected for packaging should be equally pleasing to the eye and reflect
the colour of the product contained within the package.

Colour is particularly important in self-service merchandising where
first impressions count; it has an emotional appeal and will create more
impact than any amount of reading of labels. Shape, form and design
are important, but not as important as the right colour. The colour of
the package has to do something more than reflect the colour of the
food; it also has to attract attention at point of sale, identify the brand
image, and create a sale. Fortunately, the colours of good food are excellent display colours and help to create impulse sales. The package

makes the contact with the customer and represents the brand to the public.

Because of the association between food colours and package colours, many of the considerations that apply to selection of colours for other types of packaging are less important for food packaging. The psychological appeal is more important than fashion, trends, the type of market or even the class of customer.

6.2   The function of colour

There are two principles governing the function of colour in food packaging. The first is that the package should look 'good enough to eat', and this entails an appeal to the emotions and to the subconscious psychological factors that are common to the great majority of people. The second is that the package should have visual appeal *per se* and should contribute to the selling process. Although the quality of the food is obviously the first consideration, appearance and cleanliness come a very close second. Colour is an essential part of appearance, and therefore bright and simple colours in suitable combinations will provide impulse appeal, always assuming that they are in sympathy with the colour of the food itself.

A wrong impression may be created if the colours of the package are not selected with care. The wrong shade of green on a package of peas, for example, could give the impression that the contents were stale. The use of unnatural colours in relation to anything that is put in the mouth can create antagonisms that can damage sales. For example:

- Pretty colours are appropriate to cake decorations but not to mashed potatoes.
- Purple would be acceptable for grape juice but not for gravy mix.
- Pink would be appropriate for an iced cake but not for coffee.

These typical associations are largely common sense, but there are others that are established by tradition and it would be difficult to persuade people to change their minds and to accept something different.

The really important point is that colour and food have an emotional association, and it this that makes for a successful package. Labelling, however well designed, will not achieve the desired objective alone because reading the label requires mental concentration. Colour acts on the subconscious.

6.3 Specific foodstuffs

Bakery products

The colour of packaging is important to the sale of all bakery products and particularly to flour confectionery. Good appearance is a vital selling feature and the colour of the package must enhance the appearance of the product. Iced cakes seen against a grey background would look grey themselves, and if seen against a green background might look purplish because of the effects of after-image.

Observers have noted a disappointing impression from visits to the plants of makers of flour confectionery. In some cases there has been a suggestion of poor products, and analysis usually suggests that it is the packaging that is at fault. The products looked fine until they were packed. In one case the wrappers were a dusty purple and cartons were in the same colour, with the result that the products looked 'tired' and unappetising; purple is most unsuitable for any food product because it has little impact and the wrong associations. Tests proved that the marketing of doughnuts in a blue and white package was a mistake; consumers felt that the package was too cold.

Green in the packaging of bread is to be deprecated because it suggests mould; nor is blue recommended. Red, red orange or orange would be perfectly satisfactory. Although bread is wrapped in film, or film bags, it is a pity to spoil the natural golden colour of the loaf by too much printing on the film and especially printing in the wrong colour.

The use of colour for identification purposes is helpful when bakery products are sold in a number of variations under a common brand name, such as biscuits. The brand name may be identified by a colourful and coherent house style with contrasting panels for the product description. In one example, the house style was black with a gold and white logo and the product description was carried on panels in six different supporting colours; these do not need to be food colours so long as they do not detract from the appearance of the products. This strategy helps each package in the range to sell the others by linking the products together.

Baby foods

Choice of colour for the packaging of baby foods might seem quite simple but it is, in fact, very delicate because the wrong colour can easily destroy a carefully built up brand image. Baby foods are not normally considered an impulse buy and display is not an important part of the selling process, but visual continuity is most important so that the purchaser can recognise the package that she bought on the last occasion.

Colours to be used will depend on the nature of the product. One maker uses yellow for meat products, green for vegetables and coral for fruit. Blue would be suitable for milk foods, but too pale a blue might suggest that the product was watery.

Flour

The most satisfactory way to illustrate principles for the packaging of flour is to quote a case history. A manufacturer of bagged flour was anxious to secure a greater share of the supermarket trade. As the shape of the bag could not be altered, it was felt that design and colour were the best means of achieving more consumer interest. The aim was to ensure that the package stood out better in average supermarket conditions; that it was different to competitive packs; and that it was easily recognised by the consumer. The image of the pack ought to carry through press and television advertising and it should bear a readily identifiable label.

Research indicated that the consumer wanted flour of a certain quality and she wanted the package to tell her whether the flour was self-raising and so on; she also wanted to know who packed the flour. A simple, easily recognisable package was required with a design that conveyed a clean, crisp image identifiable with flour, and it was essential that the package should stand out well on the supermarket shelf. There ought to be an appeal to the young and the design ought to include the maximum amount of white to fit in with the image of clean, white flour. Flour is not an impulse buy, and people do not stop to read what is on the package; they pick up the package on sight, and the package has to compete with the brightness of the store.

A circle was chosen for identification purposes because it had good visual impact, showed up well on the shelf, was psychologically pleasing to the average customer, and differed from the competition. The circle was coloured golden orange, the ideal hue for flour because it is the colour of fresh bread and ripe corn. Orange created the right association, had good recognition qualities and maximum impact and showed up well in the supermarket conditions. The type of flour was printed in black and white across the circle.

The name of the manufacturer was printed in a coloured panel with the name reversed in white, and this looked equally distinctive when seen in colour on television or in black and white. The colour of the panel was changed according to the type of flour, that is blue for the major line, red for self-raising flour, brown for wholemeal. These colours do not clash with orange and therefore there is no disharmony. The company symbol, a stylised version of the first letter of the company name, was printed in the same orange as the circle.

Frozen chicken

Research into the colour of packaging suitable for frozen chickens showed, rather unexpectedly, that most people said that red meant frozen to them. However, it was also found that blue made the white of the meat look fresher. One packer uses a colour coding system for weight ranges.

Fruit and vegetables

In dealing with prepackaged fruit and vegetables, the most important factor is the appearance of the product within the package; the latter is usually transparent or of such a nature that the product can be seen and examined. Tomatoes, for example, are commonly packed in an open tray and cushioned in blue paper. Green paper would be better because it has better contrast with the red of the tomatoes, but blue is traditional and well accepted by the public so there seems to be little point in altering it.

Careful research is always necessary to establish the best method of presentation in the conditions under which the produce will normally be displayed. The colour of the package or of the label is just as important as it is with other products, and it is desirable to aim for an impression of freshness; for this reason clear, bright colours are recommended.

The correct shade of green is important on labels for canned vegetables; a yellow label for a can of peas would not be acceptable but would be quite right for split peas. Research in the US indicated that the most acceptable colours for canned fruit and vegetables were:

- corn   bright gold
- peaches   darker gold
- cherries   bright red
- beetroot   dark red
- white products blue

Baked beans were found to sell better when the colour of the label was changed to a rich, dark brown emblematic of the contents of the can.

Milk

Blue is very suitable for milk and milk products because it conveys cleanliness and coolness. However, it should be a strong blue; a pale blue may suggest wateriness.

Fish

Blue is acceptable because of its association with the sea, but pink is also acceptable, especially for shellfish which is naturally pink.

Cake mixes

Different flavours will sell better if the colours of the package are also different. A cake mix packed in a black carton was a total failure although it stood out well on the shelf.

Soup mixes

Dehydrated soup mixes are particularly difficult to pack and have to be checked for suitability on packaging machinery. They tend to take up oxygen from the air and must be tested under different conditions of light and temperature. Sachets are usually the best method, contained in a carton and shrink wrapped, but this method may limit the use of colour.

Salad dressings

Different colours have been found to inspire customers to try different types of dressing.

Coffee

The colour of the can has been found to affect the taste in the eyes of the customer. Mauve has been found to have a particularly adverse effect.

Tea

Research into falling tea sales in the US indicated that tea, in that country, was thought to be 'cissy' and only suitable for old ladies and invalids. As part of a campaign to alter that image, the colours used for packaging were strengthened and the use of more vigorous colours proved very successful.

In this country there is less need for such a strategy. However, the market is very competitive and the package should have maximum impact at point of sale; it must stand out well in supermarket conditions and it must be easily recognisable. So far as is known there are no colours which have any particular association with tea, and colour can be used to create maximum impact on display or to create a flavour or strength image. A tea package must have a recognisable visual image because brand loyalties are very strong and the package must compare well with the market leader. The design on one side of the package should be horizontal, and on the other side it should be vertical, so that the package can be read or recognised whichever way it is displayed on the shelf. The design on the end of the package should not be forgotten.

The design should be uncluttered; it should have strong, clear lettering and there should be good contrast between background and lettering. If the tea is portrayed, it should be hot, steaming and strong. Some distinctive style for the brand name is recommended and there should be a link between various types of tea from the same source. Colour may be used to distinguish between different blends; for example, a delicate green would be very suitable for a tender leaf tea and red for a strong, well flavoured tea. It is very desirable that colour should be maintained to strict standards lest consumers should feel that some change has been made in the blend.

6.4 Applying colour

Packaging colour is not only the colour on the carton or label, but may also be the colour of the container itself. In the latter case the effect of the container colour on the appearance of the food within must be considered. White is always correct for a container but it is 'ordinary'; the additional appeal of other colours is worth thinking about, and it is easy enough to obtain any desired colour in plastics. However, a deeply coloured container may alter the appearance of the contents by reflection, while after-image can make the food look less than appetising when the container is opened. Careful testing may be necessary to establish the best effect.

The insides of cartons or boxes are equally important. White is always safe, but if other colours are used they must be selected with care to enhance the appearance of the product. Most British people do not like drinking tea out of coloured cups, and they may not like eating food out of coloured containers in those cases where the food is consumed directly from the container. In all cases, check the appearance of the food when the package is opened.

There are a number of points that require special attention when applying colour to food packaging, for example:

- Ensure good proportions of colour and avoid combinations that might cause after-image effects.
- Colours that are too hard or bright may be a mistake; they glue the eye to the surface of the package and suggest lack of shelf life.
- The colour of the package sets up expectations about the flavour of the food and may influence the actual taste; people judge the taste and aroma of the product to some extent by the colour of the package. Make sure that the colours used have the right association.
- Colour reproduction is important; poor reproduction may change the image of the product.

- Draw a distinction between the colour *on* the package and the colour *of* the package.

6.5   The right colour

A good deal of research has been devoted to the subject of colour and food. There is an art in selecting the right colour to merchandise food-stuffs; the same direct associations with, and known responses to, the stimulation of colour can be applied to food packaging. Bright and warm colours (red, orange, yellow) tend to stimulate the autonomic nervous system, which includes the digestion, while soft and cool colours (blue, green) retard it. It is no accident, therefore, that the best colours for food packaging are red, orange and yellow, the colours of good food itself.

The common pattern which is revealed by the study of the appeal of colour to the appetite can be charted, and the curve showing the average responses of ordinary people reaches a high point at the orange and red orange sections of the spectrum. This includes the colour of fresh bread, oranges, and other good things to eat, all of which are well liked and appetising. The curve reaches its lowest point at yellow green and purple, which are the colours of sickness and bad meat respectively. Green, the colour of fresh vegetables, is well liked and so is blue because it is clean and cool, although there are few blue foodstuffs. Blue is best used as a background to food, and blue green is an excellent background to meat because it is the complement of the red of the meat.

These typical likes and dislikes are psychological in origin and stem partly from the autonomic responses mentioned above and partly from intuitive associations which have built up over the years. They can readily be verified by observation and by the application of common sense. Those colours that are good or bad for food are equally good or bad for food packaging. The colours of the package should reflect and enhance the product within the package.

Extensive research has shown that the best colours to use for food packaging are, in order of preference:

• red	the best of all colours: high attraction value
• red orange	equally good and with even higher attraction value
• orange	good attraction value but not so strong as red orange
• peach	good background for many foodstuffs: appetising

- pink   lacks strong attraction at point of sale but recommended for sweet things
- tan   excellent for bakery products
- brown   warm shades of brown have an association with many foods
- yellow   highly visible and recommended to attract attention
- green   well recommended but neutral in nature
- white   always acceptable but lacks impact
- blue   essentially a background or foil colour
- blue green   a background or foil colour, especially for meat products
- purple   not recommended
- lilac   paler shades of violet may be used for some bakery products
- grey   not recommended for any foodstuff packaging
- black   contrast only

The actual hues to be used in a particular situation will depend on a number of factors, the most important of which is the nature of the product.

6.6   Factors affecting selection

The product

The nature of the product, and the colour of the product, are the key factors in the selection of colour for food packaging. The colour of the package should have the right association with the product and with what people believe to be the right colour for the product; this applies particularly to those products that have a well established and unchanging colour, such as butter or meat. There are specific associations with products and these can be identified and used with advantage; certain colours are traditional and there is nothing to be gained by altering them.

Common sense, observation and research should be applied to this problem; the more obvious associations such as red for meat, yellow for butter, and so on are common sense. Some others are listed in the following section. In other cases, discover what people generally associate with the product. The colour of canned peas, repeated on the label, would do nothing for sales; the majority of people associate bright green with peas and the colour of the contents of a can of peas are anything but bright.

Product ranges

This factor is of especial significance in food packaging because the same product may be sold in a number of different flavours or varieties, and the same manufacturer may sell a number of different products under a corporate brand name. The design or logo indicating the corporate brand name or image need not be in food colours, but the colour used for product identification should be appropriate to that product. Different flavours or varieties of the same product can be identified by appropriate colour. For example:

- Farex used a green package for baby food but their rice cereal was distinguished by a yellow orange band.
- Boots's foods are colour coded to distinguish diabetic, dietetic and conventional foods.
- An American canner uses the background colour of the label to distinguish the class of product, for example ruby red for tomatoes and fruit cocktails, golden yellow for peaches and corn, green for beans and other vegetables, soft blue for colourless fruit and vegetables, apple sauce, and so on.
- A producer of vacuum packed meats uses a four-colour label for all products. Red, white and black are common to all varieties, but the fourth colour identifies the contents, for example brown for roast pork, dark blue for silverside, green for savoury loaf, light blue for ham.
- Golden Wonder use colour coded packages to distinguish varieties. Yellow and red are the corporate colours, with blue for the logo; varieties are distinguished by green, brown, magenta and mauve.
- Bird's Eye introduced a new strategy for its ready prepared meals, including new packaging, with blue packages for fish meals and red for meat meals, although all packages have a uniform design on the front. They also group together similar lines, such as by using 'Country' as a collective term for vegetables. These ideas are claimed to make it easier to spot Bird's Eye in cabinets.

Product features

The association between colour and taste is important. Many people are unable to judge the flavour of a product if the label does not give them some clue; expectations from the colour of the package which are not borne out by actual taste may cause dissatisfaction, and testing may be necessary in case of doubt. The colour of the package may emphasise flavour. Pastels generally suggest sweetness, and strength of colour may suggest strength of flavour; it may be necessary to tone down the colour to suggest subtlety of flavour.

The colour of the package should associate with the smell of the product where this is significant, and the colour may be used to emphasise aroma.

The health benefit of food is becoming an important selling point. Tesco have introduced colour coded symbols to indicate the health qualities of their own label foods, for example:

- dark green   high in vitamin C
- light green   rich in vitamin C
- deep yellow   low fat
- pale yellow   no added sugar
- red (five   low calories
  variations)   vitamin fortified
                      high in vitamin A
                      low sugar
                      high fibre
- blue (two   low salt
  variations)   no added salt
- orange   reduced calories

Some manufacturers are using white to emphasise health aspects. However, all health foods cannot come in white packs, and it is better to make the food look appetisingly beautiful rather than medically dietetic.

Product image

Where there is a product brand image, the brand colour should be selected because it is suitable for the food, but different flavours or varieties may be distinguished by appropriate coloured panels. This, and corporate brand image, are discussed above.

Type of packaging

This will obviously depend on the nature of the product, but plastics containers require a different approach to colour selection. They have many advantages in relation to foods such an ease of handling, light weight, fewer breakages, security, re-use value, and something different in design. The plastics used are mainly polystyrene and polypropylene, the latter being suitable for products with a high fat content. The main problems are to decide whether to use a white or a coloured container and whether to print on the containers. Comparatively little use has been made of coloured containers, for economic reasons, but they do provide something 'different' and give the designer more scope with colour combinations. Clear or translucent plastics containers are an alternative to glass.

Coloured film might be used for overwrapping more expensive products and provides additional attraction. Clear film is preferred for meat products.

The market

The market is unlikely to be of great significance in selecting colours for food packaging; butter is yellow whatever the nature of the market. Broad principles suggest bright and simple colours for the mass market, but more sophisticated products will gain from subtle variations. This principle is seen at work when butter is wrapped in gold foil.

For export markets the whole packaging may require rethinking. The sales appeal of the product may have different features in different markets and countries. A convenience food in the UK may sell on vitamin content, but elsewhere it may sell on some other feature. Different packages, different designs and different colours may be called for. Most countries associate yellow with food products, but in Switzerland it conveys cosmetics; the French associate green with cosmetics, but the Americans associate it with confectionery.

The customer

The nature of the customer has little significance in selecting colours for food packaging; all types of customers will be influenced by the same factors. Certain specialised markets may require individual thought, such as baby foods. Some items may be aimed specifically at the teenage market and would benefit from colours that appeal to teenagers. The health food market has an AB following and this suggests more sophisticated colours.

Purchasing habits

These are significant when the package has a function in the home, probably in the kitchen. The package may even be designed specifically for home use, and it may be desirable to ensure that the colour of the package is compatible with the average kitchen.

Selling conditions

Display is, of course, vital to food sales and fortunately most of the colours which are most suitable for use with foods are excellent for display purposes. A typical supermarket stocks some thousands of items and therefore any one package has to compete against competition of a very high order. Many competing packs will be using more or less the same colours and consequently there is a premium on good design and appealing combinations of colours; exact shades may be especially important. Where the package is part of a range, say of different flavours, it may be possible to use colour to create excitement at point of sale by grouping the range. Anything that can be done to attract attention will help sales.

Distribution

Distribution has little significance in colour selection, except that a food package intended for the catering trade might have different colours to one intended for the consumer trade.

Competition

Competition is very intense in the food field because of the nature of the grocery trade and the merchandising techniques that are employed. It is vital to study the colours of the competition and how they are used. It may be difficult to improve on the colours used by the competition because all manufacturers may use the same reasoning. Variations of shade may have to be used to establish a difference, and colour combinations may have to be manipulated in a sophisticated way. In some cases it may be necessary to abandon established principles in order to achieve something startling.

Method of promotion

The sales theme may influence selection of colour when it is desired to convey a particular image, say a luxury image. It would be best to use richer variations of the basic hues rather than an unattractive colour.

Technical features

If the container has a re-use value, such as spice jars, then the colour of the closure might follow kitchen colour trends. Closures are the one sector in food packaging where trends can have priority. The colour of the closure may also be used to identify flavours, varieties and so on.

Labelling

Health labelling may be important in the future. Tesco have recently introduced a nutrition panel on their own label foods, combined with colour coded symbols (mentioned above).

Product appearance

The package colours must always enhance the appearance of the product, especially when the package is opened for the first time.

6.7   Colour applications

In most food packaging applications, strong and vivid colours will be required because of the need to achieve impulse attraction at point of sale. However, pastels may be used to suggest a more sophisticated or expensive product and where it is desired to convey softness or gentleness.

It would be desirable to pay some attention to fashion and to select those variations of basic hues which accord with current trends of consumer preference for colour, as long as this does not conflict with the functional application of the hue.

The colour analysis that follows has been set out in some detail because of the importance of the subject, but it is summarised in the colour index in Part V.

- hard colours   generally preferred for most food packaging because of their impact and because they are inviting to the viewer
- soft colours   generally best for background but may be used to suggest sophistication: pastels suggest sweetness
- bright colours   preferred for most food packaging applications because of the display element: have more emotional appeal
- deep colours   not recommended except to suggest strong flavour; also suggest stability and weight

Violet group

Association with food   Avoid violet shades in association with food, although mauve and lilac may be used for sweet things. It is particularly unsuitable for coffee, but is the natural colour of grapes and grape juice.

Food packaging   Purple and violet are not recommended but lilac and similar pale shades might be used for some bakery products. Cadbury purple is an exception to most rules.

Products   Some bakery products, grape products.

Smell   Lilac, violet, lavender.

Taste   No application.

Blue group

Association with food   Does not have any direct association with foodstuffs but is an excellent foil. Cool, clean and well liked. Blue tends to suggest a mild taste and pale blue may suggest that a product is watery.

Food packaging   Blue is essentially a background or foil colour, but denotes coolness or cleanliness. Cambridge blue has been used for baby foods and white vegetables. Suggests sea and water and is suitable for sea foods.

Products   Sea foods, milk, dairy products, baby foods, tomatoes, white vegetables, frozen chicken.

Taste   Mild taste.

Avoid   For packs in Sweden. Not for bread or bakery products.

Blue green group

Association with food   Blue greens do not have any direct association with food. However, they are a good foil, especially for meat because blue green is the direct complement of red and helps to make meat look more appetising. Feminine colour.

Food packaging   As for blue but is especially good for meat products.

Products   Meat, otherwise as for blue.

Taste   Mild taste.

Avoid   Use with bread.

Green group

Association with food   The natural colour of vegetables, fruits and other good things to eat; select the shade of green carefully to reflect the colour of the food. Green should be avoided in association with meat because it suggests spoilage, and also in association with bread because it suggests mould. Olive is best avoided except where it is the natural colour of a product. Marked yellow greens and greenish yellows are best avoided because they suggest sickness, but lime green is the natural colour of many products and has good attraction value. For Easter.

Food packaging   For vegetables and country products generally, but select the shade with care to convey the right association with the product. Lime green and similar shades have good attraction value, but otherwise green tends to be neutral. Green conveys cosmetics in France but confectionery in the USA.

Products   Vegetables, peas, country products, baby foods.

Smell   Pine, balsam, winter green, sage, olive, apple, peppermint.

Taste   Lime.

Avoid   Use with bread, meat or iced cakes. Yellow greens, greenish yellows.

Yellow group

Association with food   The natural colour of butter, cheese and other eatables. Excellent for food, especially in paler shades. Gold has a high class image. Stimulates the appetite and has good attraction value. For children, for Easter.

Food packaging   Highly visible and recommended to attract attention, but use 'butter' yellow and not harsh variations. Excellent for appeal to children. Conveys food in most countries except Switzerland, where it conveys cosmetics.

Products   Butter, cheese, baby foods, split peas, corn.

Smell   Vanilla, lemon, honeysuckle, saffron.

Taste   Lemon, mild flavour.

Avoid   Mustard tones, harsh yellows, greenish yellows. Do not use with peas.

Orange group

Association with food   The colour of fresh bread, oranges and other appetite provoking foods. Red orange is one of the best of all colours for food; it attracts attention and creates appetite appeal. Particularly suitable for bakery products.

Food packaging   Red orange is well recommended for many foodstuffs, particularly those with an orange flavour. Paler shades are equally suitable but are less powerful. May be a little overpowering in large areas.

Products   Bakery products, bread, flour, cereals, meat, baby foods.

Smell   Orange, apricot, tangerine.

Taste   Orange.

Avoid   Too large areas.

Brown group

Association with food   The natural colour of many foodstuffs, such as coffee, chocolate. Warmer shades such as tan can be recommended for many products, but avoid earthy browns. Brown bread is well liked and brown shelled eggs are preferred in the UK.

Food packaging   Relate the shade to the actual colour of the product, but avoid earthy browns which might suggest dirt.

Products   Baked beans, nuts, coffee, bakery products, chocolate, corn.

Smell   Coffee, cedar, chocolate, chestnut, cinnamon, ginger, nutmeg, almond.

Taste   Cola, strong beer, coffee. Brown ice cream is considered to be chocolate flavoured.

Avoid   Earthy browns.

Red group

Association with food   The natural colour of meat and one of the best colours for food; stimulates the appetite and has good attraction value. A friendly colour although some variations are repelling: warmer shades are recommended.

Food packaging   Warm shades are preferred and have high attraction value, but red orange is often better for maximum impact. Variations such as flame, scarlet or coral can be recommended. May be a little overpowering in large areas.

Products   Meat, meat products, cherries, beetroot, bread, baby foods (particularly coral), frozen chicken.

Smell   Plum.

Taste   Conveys a rich flavour in some cases but a harsh flavour in others. Strawberry.

Avoid   Purplish reds which may suggest bad meat.

Pink group

Association with food   Peach is the colour of many fruits including the peach itself, and is well recommended for many foods especially confectionery and sweet things. Appeals to children. Pink is appropriate for an iced cake but not for coffee.

Food packaging   Peach is suitable for many fruits and is good background for many foodstuffs. Pinks are recommended for sweet things, including confectionery, but lack impact at point of sale.

Products   Fruit, confectionery, sea foods.

Smell   Carnation, rose, peach.

Taste   Sweet things generally.

White

Association with food   White is always correct in association with food and conveys cleanliness, but it is best used for background. Lacks impulse but goes with any other colour.

Food packaging   Always correct but lacks impact. Excellent as background and is the universal alternative. Conveys purity and has been used for health foods to emphasise the health aspect.

Products   Any.

Taste   White ice cream is considered to be vanilla flavoured.

Avoid   White and blue together for bakery products.

Off white group

Association with food   Best avoided in association with food because it may suggest dirt, but warmer off whites may be used in suitable circumstances and cream is always correct.

Food packaging Warmer off whites might be used as background to suggest softness and quality.

Grey group

Association with food Avoid in association with food; grey bread or grey meat would be rejected, and so would bacon on a grey plate.

Food packaging Not recommended, especially in association with iced cakes.

Black

Association with food Avoid in association with food, although it is acceptable in some circumstances, such as black bread, black puddings.

Food packaging Few uses except to create contrast. Black tops or closures may be used to make a package stand out. Not recommended for cake mixes but acceptable for tea.

7   Industrial products

Packaging is equally important to the marketing of industrial products, although there may be different emphasis on the various functions. Ensuring that a complex piece of machinery reaches its destination in good condition is part of good marketing because it ensures a satisfied customer, but the package does not sell the industrial product in the way that a package sells an Easter egg.

Customer satisfaction is just as important to the sale of raw materials as it is to the sale of finished products. The sales appeal of a raw material may lie in sacks which can be easily handled on a production line, and it may be necessary to redesign the sacks to meet the needs of a new method of production.

A good looking, well designed and functional package will add to the image of any product whether it is industrial or consumer; a cheap package conveys an impression of a cheap product.

The key factors in designing packaging for industrial products are protection, convenience and identification. Protection needs no special comment. Convenience and identification may be of primary importance, particularly with replacement items and spares which may have to be stored for long periods and found in a hurry. Some packages may be carried around by maintenance engineers and the like, and colour is very useful as a means of identifying size, thread and other characteristics.

The function of the package will be dictated by the way that the product is used and the quantities required at any one time. Careful research into the needs of users will pay dividends. Kits, and the contents of kits, is another area that will justify research; a spare part without its appropriate fastenings is useless.

Many so-called industrial products have a consumer or semi-consumer sale. The considerations that apply to consumer packaging

219

generally will apply equally to such products, with the added possibility that more protection will be required because of the industrial use.

The following are some miscellanous comments:

Chemicals   Neutral colours are normal except for household products, but especial care is necessary with the labelling of hazardous chemicals and those which contain notifiable constituents. EEC hazard labelling regulations are strict and mandatory, and if a hazard label is needed the text must be exactly as in the legislation.

Electricity   Blue is associated with electrical power.

Farmers   Green is a poor colour for products sold to farmers because the farmer is in a green atmosphere all day. Red is better.

Machinery   Red is a traditional colour, particularly for exhibition purposes, and can be used for packaging.

Steel   Blue is traditional for steel products.

8 Other products

8.1 Beverages

The broad function of a beverage label is to identify the brand readily; to stand out well in artificial and natural light; to convey the right association with the product or flavour; to attract attention to itself; and to create an up to date image without losing brand identity.

Whisky The total effect should be masculine and aggressive and it must look impressive in a mass display, such as at an airport.

Soft drinks Primarily a teenage market and can have a modern image. Frequently stored in a fridge, and labels should stand up to this.
PET bottles are now common; bases provide opportunities for identification.
Bands, or illustrations, in different colours will help to identify flavours.
Bitter lemon and other chasers are more sophisticated drinks and benefit from a quality image.
Brown identifies cola drinks. Green identifies lime drinks. Yellow identifies lemon drinks, although lemonade is brown or white in different parts of the country.

Wine Wine is liked in pink, golden yellow and claret red, and these colours are suitable for labels. A neutral label may, however, be better for a vintage wine so that purchasers can judge the quality of the wine itself. There is a thought to be potential for wine in cans, and 'bag in box' packages are now common.

Beer Brown bottles are common to counteract the effect of light on beer, but bottled beer sales are declining and people seem to prefer cans. No special comments seem necessary about the colour of labels.

221

8.2 Chemists' products

For the purposes of this book, this heading does not include cosmetics (discussed in Section 1 of this part) or baby foods (discussed in Section 5 of this part).

Baby products Baby blue and baby pink are traditional but lack impact in packaging applications. Red, yellow and strong blue would be better.

Soap Red and brown are traditional for carbolic soap, but brown also denotes laundry soap. (See also cosmetics.)

Pharmaceuticals White, black or neutral are prescribed for ethicals. Restrained colours are recommended for other products with a suggestion of a clinical image. People welcome foil strips (for pills and so on) but complain about the difficulty of opening them. All bottles must have child proof closures.

Sanitary towels Sell very well in white packs, but other colours that have been used successfully are light blue, dark blue, pink, mauve, pale green. Colour is often used to identify qualities.

Toothpaste Avoid yellow green which is not liked for things put in the mouth.

8.3 Household products

For most household products use compelling colours with strong impulse attraction. The variation will depend on the product but may have to be suitable for kitchen or bathroom use.

Detergents Plastics bottles are generally acceptable but should look safe to handle. A toilet bowl cleanser should have a package that will not look out of place in the bathroom.

Kitchen products Purple has been found to be unsuitable. Consider kitchen colour trends.

8.4 Miscellaneous products

Motor oil The sales theme plays a part in the selection of colours for packaging. At one time, Castrol used a sales theme based on fear of breakdown and used a green can with the logo in red; green is the British racing colour and has the right association with the product. Black was eliminated as a can colour because it had the wrong associations with the product, including dirt, death and accidents. Blue was acceptable because it was an excellent foil, universally popular and contrasted well with the green of the can; it also had the right associations with the 'fear of breakdown' theme and was used for point of sale display. Later, Castrol adopted a new sales theme based on 'clean oil' and green for the cans was abandoned in favour of white.

Photography   Kodak yellow is associated with photography virtually world wide.

Typewriter ribbons   Coloured cellophane has been used to wrap ribbons and to identify the various types. Colours used were red, chocolate, mauve, green, blue, pink.

8.5   Cigarettes

For health reasons many packages now use white because this is associated with purity and cleanliness. Green is used for menthol cigarettes because it suggests coolness. Red tends to suggest a harsh taste but is used for quality cigarettes in association with gold. Pink is feminine and has been used for cigarettes aimed at women. Brown suggests maturity. Black is a little funereal but has been used with success by Player; it is 'different'.

Trade marks should be shown on every cigarette in African countries.

Part V
COLOUR SELECTION

1 The approach

Colour is a vital part of a successful package provided that the right colour is selected, and this part is concerned with the method of selecting the right hue or variation of hue. The process can be complex and requires a great deal of thought and consideration; it is recommended that it should be approached in a systematic way and that guesswork and personal prejudices should be avoided as far as possible.

The first step in effective selection of colour is to analyse the job that the package has to do and to isolate the various functions that colour can, and should, perform if the package is to do its job properly. It is necessary to identify those attributes and characteristics of colour which are most appropriate to these functions, which may include attracting attention, creating impulse, ensuring a satisfactory image, appealing to a specific type of customer, and the other points listed in the design brief, not forgetting the right association with the product contained in the package. In other words, the first step is to decide what colour has to do to produce the ideal package for the situation under consideration.

Although there may be many hues which have characteristics appropriate to some of the functions, there will be very few which have attributes appropriate to all the functions in a specific case. There may be only one such hue – and that is the one to select for the package.

Research and experience has shown that there are a number of hues which are particularly suitable for packaging (and some which should be avoided), and these should have priority in any selection. However, it is still necessary to identify the hue, or hues, which have the most appropriate characteristics for the situation under consideration. This can be effected by matching the functions that colour has to perform to the characteristics of individual hues.

The recommended method is to prepare a colour specification for each individual package, or series of packages with a common theme or link, based on the design brief outlined in Part III. The specification should set out the various functions that colour has to perform and the attributes and characteristics of colour which will be required to achieve these functions. The various characteristics of colour which are appropriate to packaging are listed in Part II and will help in the compilation of the specification. The ideal colours for packaging are listed in Section 2 of this part and the two are brought together in the colour index which lists the characteristics of individual hues and the job that they can do (Section 4 of this part).

The hue to select is that one which has the most attributes matching the needs identified in the specification.

It should be made clear, however, that there can be no hard and fast rules where colour is concerned, and selection should never be made on a purely mechanical basis. Colour is an emotional thing, and the most unusual colours or combinations of colours will often be successful, but they should only be adopted after careful consideration of all the circumstances and with a clear view of what has to be achieved.

2 The packaging palette

The broad rules of colour usage in packaging applications are dictated by the basic physiological and psychological reactions to colour, for example:

- Hard colours have more impact than soft ones.
- Bright colours have more impact than dark ones.
- Pure colours are preferred to greyish ones.

Using these basic principles it is possible to suggest a short range of colours which are particularly suitable for packaging and which have excellent visibility and emotional appeal. They are particularly suitable for creating impulse sales because all people are attracted to them subconsciously. These are known as feature colours.

Feature colours

Hard colours

- red — clear, warm red
- red orange — better than pure orange
- yellow — clear, warm yellow

These are pure colours, dynamic, and pleasing. The first two have high impulse attraction; yellow has rather less impulse attraction but it is luminous and has high visibility.

Soft colours

- green — bright, apple green
- blue green — particularly suitable for food
- blue — clear, deep blue

These are also pure colours and well liked. They are less dynamic than the hard colours and are best used as background, although green can also be used as a feature colour. The ideal is to use hard colours on a soft background, but this may be modified where necessary.

Pastel variations of these colours can also be used if they have sufficient brightness; it is recommended that the variation selected should be halfway between pure hue and white.

Foil colours

- white
- black
- maroon
- dark green
- midnight blue

These foil colours contrast with the feature colours and set them off to best advantage. If the hard colours are used for the principal features of a design, or as a background, the foil colours can be recommended for secondary items.

Secondary colours

- orange   may be used, particularly for foodstuffs, but red orange has more impact.
- brown   certain shades of brown may be used with advantage where appropriate, but generally brown lacks impact and is not well liked

Modified colours

- yellow orange
- mustard
- orchid
- lilac

Modified colours are best avoided in packaging applications because they are not clearly perceived by the eye and are not very easily noticed. They may, however, be used in circumstances where sophistication or a degree of fashion is required.

Colours to avoid

- yellow green   not well liked; tends to suggest nausea, particularly in association with food; lime green may, however, be used in suitable circumstances
- magenta   weak in motivation; not recommended

- violet   weak in motivation; not recommended
- grey   lacks impact in packaging applications
- olive green   deep colour which is not well liked
- purple   deep colour, not well liked

The feature colours and foil colours listed above have been proved by research and market experience, but they are not the only colours that can be used with advantage. The hue to be selected in a specific case, and the variation of that hue, will depend on the characteristics of colour required to meet the colour specification.

The characteristics of colour which are appropriate to packaging applications are listed in Part II and the characteristics of each hue are set out in the colour index in Section 4 of this part.

3 The colour specification

3.1 The preliminaries

Before attempting to formulate a specification, it will be useful to consider the marketing plan in broad terms. Packaging is an essential part of the marketing mix and all concerned need to be quite clear about what they are doing and how colour can be used to maximum advantage; agreement should be reached on contentious points.

It is assumed that there is a positive sales plan and that elements of a design brief exist, at least in embryo. Some points that will benefit from consideration include the following:

- Policy with regard to product ranges and how colour is to be used to differentiate them.
- The function of the package in relation to corporate image and brand image and the part that colour is to play.
- What the package is intended to accomplish. For example, it may be intended to reach a new market; colour should be aimed at that market.
- What colour has to do. The most important function of colour is usually to attract attention at point of sale. However, the colours of an individual package may also have to appeal to customers of a certain age group, be appropriate to the product, convey an image of luxury, support a specific sales theme, create a mood or emotion and create a brand or corporate image.
- The image that it is desired to put over. Images might include:
  hard sell, such as for impulse sales
  soft sell, such as for a prestige product
  flashy
  sophisticated

    pleasurable
    new and up to date
    traditional.

- How colour is to be achieved. Materials to be used and methods of reproduction may impose restraints and some changes may be necessary.
- A review of the characteristics of colour listed in Part II and how they might be used to best advantage. They may be used to emphasise any feature of the product that it is desired to underline or to enhance the features of the product most likely to appeal to customers.
- How co-ordination is to be achieved between the package and other forms of promotion, particularly television advertising, and how colour can assist.
- Whether illustration is to be used and its implications from a colour point of view.
- The sales theme to be used and whether colour can support it.
- The requirements of those specific products listed in Part IV.

3.2 Preparation

Preparation of a colour specification for each individual situation should be based on the design brief set out in Part III. Each item of the brief, where colour is relevant, should come under review and be listed in the specification.

- Where possible, define the factors affecting colour choice exactly, for example the colour of the product itself.
- Where there are no positive factors to be defined but where colour has a part to play, consider how it might be used to best advantage. For example, consider the type of market and whether colour can be aimed at a specific market.
- In all other cases review the possibilities of colour, taking a common sense view.
- Whenever possible list the colour characteristics required.

The following is an outline specification. The numbers in brackets refer to sections of the design brief in Part III.

	action	*colour characteristics (see Part II)*
The product (2)		
Marketing aspects (2.1)		
the product	define product colour and enhancing colours	associations
the package	review	
nature of sale	review	
	impulse purchase	impulse colours
	considered purchase	fashion colours.
selling features	define	as appropriate: smell, taste, etc.
repurchase	review	
price	define	
	low price	bright colours
	high price	sophisticated colours
product associations	define	product associations
Product image (2.2)		
product identity	define	as appropriate
corporate identity	consider	special consideration
brand identity	consider	special consideration
Physical characteristics (2.3)		
adverse characteristics	define	as appropriate
temperature range	consider	temperature control
rancidity	consider	protection
toxicity	review	
processing	review	
light	review	
safety	review	contrasts
product appearance	consider	
Types of packaging (2.4)	define implications	
Package size (2.5)	consider	size
The market (3)		
Nature of market (3.1)	define	markets
Most profitable market (3.2)	consider	
Type of market (3.3)	consider	
Speciality markets (3.4)	consider	seasons
Export markets (3.5)	consider	see Part III.
The customer (4)		
Type of customer (4.1)	define	as appropriate

	action	*colour characteristics*
Age of customer (4.2)	define	age
Sex of customer (4.3)	define	sex
Purchasing influence (4.4)	consider	as appropriate

Purchasing habits (5)

Buying habits (5.1)	review	
Reasons for purchase (5.3)	review	
Motivation of purchaser (5.4)	consider	mood
Shape and size of package (5.6)	consider	size
Circumstances of purchase (5.8)	consider	
Impulse purchase (5.9)	consider	impulse

Usage of the product (6)

The package (6.2)	review	
Storage (6.4)	review	
Where used (6.5)	consider	trends
Suggestions for use (6.6)	consider	readability

Selling conditions (7)

Type of outlet (7.1)	review	
Method of display (7.2)	review	
Shelf display (7.3)	review	
Lighting	review	

Distribution conditions (8)

Storage conditions (8.2)	review	
Information (8.4)	consider	readability

The retailer (9)

Regulations (9.3)	consider	readability
Up to date (9.6)	consider	trends
Display (9.8)	review	

Competition (10)

What is the competition? (10.1)	review	
Who is the competition? (10.2)	review	
Advantages (10.4)	review	

	action	*colour characteristics*
Method of promotion (11)		
Advertising generally (11.1)	review co-ordination	
Television advertising (11.2)	consider	television
Special offers (11.3)	consider	impulse
Brand promotion (11.4)	review	
Special sales themes (11.5)	define	as appropriate
Seasonal sales (11.6)	consider	seasons
Gift sales (11.7)	review	
Mood (11.8)	consider	mood
Limitations on design (12)		
Machinery available (12.1)	review	
Availability of materials (12.3)	review	
Economy of materials (12.4)	review	
Production tolerances (12.5)	review	
Different materials (12.6)	review	
Cost (12.7)	review	
Materials used (12.9)	review	
Shape (12.10)	consider	shape
Technical features (13)		
Form, fill, seal (13.2)	review	combinations
Mechanical filling (13.8)	review	
Finish (13.11)	review	
Reproduction (14)		
The substrate (14.1)	review	
Method (14.3)	review	
Labels (14.4)	review	
Colour (14.6)	review different colours	
Inks (14.7)	review	
Quality control (14.10)	review	
Labelling (15)		
Product regulations (15.1)	consider	readability
Hazards (15.3)	review	
Storage (15.4)	consider	readability
Cooking (15.5)	consider	readability
Price (15.6)	review	
Bar coding (15.8)	review	
Product content (15.9)	review	

	action	*colour characteristics*
Instructions for use (15.11)	review	
Export (16)	consider	see Part III.
Appeal to the senses (17)		
Customer prejudices (17.1)	consider	associations
Product appearance (17.3)	review	
Graphical aspects (18)		
Illustration (18.1)	review	
Identification (18.2)	consider	coding
Legibility (18.3)	consider	readability, recognition, visibility
Supplementary packaging (19)		
Caps and closures (19.1)	consider	see colour index
Plastics bottles (19.2)	consider	product
Wrappings (19.3)	review	

Note   Colour is particularly important in relation to foodstuffs and attention is drawn to the notes in Part IV.

3.3   Colour selection

Some of the items in the specification may require study in considerable detail to determine the part that colour can best play and which colour characteristics will be most appropriate. A number of contentious and conflicting arguments may develop, but due weight should always be given to the primary purpose of the package and to the part that colour can perform in achieving the ideal package. The end result should be a list of colour characteristics appropriate to the objective of the exercise.

When the specification is complete, colour selection can begin. There is no automatic way of making a final selection; nor should the process be mechanical. Skill, experience and common sense should be applied if the best results are to be achieved.

The suggested steps in selection are as follows:

• With the list of colour characteristics derived from the specification, and with the graphic design in mind, look at the colour index in Section 4 in this part and pick out those basic hues which have the appropriate characteristics.

- Decide whether hard or soft colours are required. Generally speaking, hard colours have more impact and are best for packaging applications; they should be used for feature colours. Soft colours are used for background. However, there is no hard and fast rule about this and soft colours may suit the marketing strategy better, although it is seldom a good idea to use a soft feature on a hard background. A good general rule is that elements in a design which are intended to attract attention should be hard and set against a soft background.

  If the package design features a symbol or logo, the symbol and the secondary colour should almost always be hard.

- Decide which is to be the feature colour and which the background or foil colour, and what other colours are required to make up the graphic design. All the colours used should, as far as possible, have some of the specified characteristics but it is equally important that the colours should enhance, or contrast with, each other harmoniously.

  Colours should be kept as simple as possible; the eye prefers simplicity of colour just as it does simplicity of shape.

- Review the packaging palette in Section 2 of this part and let the recommendations made therein influence decisions. This palette reflects much research and experience.

- When the main feature colour has been selected it may be left to the designer to select secondary colours which combine well with the feature colour, but it is better if the secondary colours have at least some of the specified characteristics.

  There may be only one hue which meets all the criteria – this would normally be the feature colour – and in that event secondary colours may be selected because they emphasise the feature colours.

- Decide whether light or dark variations are best. In most cases a medium tone is recommended for packaging, but light variations, including pastels, may be used when product or market so dictate. Very dark variations are best used as foils, that is as contrasts or for secondary items. As a general rule, use brightness against darkness.

  A colour having a comparatively low reflectance value will be accentuated by a colour of high reflectance value, for example a medium red on a background of pale blue. The important point is to ensure that colour stands out in average selling conditions.

- Decide whether pure hues or modified hues are desirable. Mass market products generally benefit from strong primary colours but luxury products, and those aimed at specialised markets, may require something more subtle and sophisticated. Modified colours are mixtures of adjacent basic hues and do not have as much impact as the basic hues themselves, but they may be used in appropriate cases, as may variations of primary hues which reflect current vogues.

● Decide whether there is a case for muted (greyed) hues. Most packages benefit from bright, clear colours because these have most impact, but muted versions are often used to convey fashion especially when these muted versions are popular trend colours. For example, if it is decided that blue is the most suitable feature colour, use a strong, clear blue as a general rule especially for mass market products; however, a product having a fashion connotation might benefit from a pale, smoky blue depending on trends of consumer preference for colour at the time.

As a general rule, use pure colours against greyed ones.

● Consider trends: see Part II, Section 8.

● Consider colour usage: see Part II, Section 9.

● Consider any modifications which may be necessary for functional reasons, for example protection against light, limitations of printing inks and the light fast qualities of inks. See Section 3.5 in this part.

● Revise, refine and compromise where necessary to obtain the best result.

3.4 Usage of colour

The way that colour is used in graphic design and the way that the colours are combined together is largely a matter for the skill of the designer, as explained in Part II Section 9, but particular attention should be paid to the following:

Colour modifiers There are a number of phenomena which can make all the difference between an attractive package and one that fails in its purpose. Although most of these will be known to skilled designers, they are worth checking.

Colour combinations Although colours can be combined in various ways and according to academic rules, the important point in packaging is that the combinations should be emotionally acceptable and, in appropriate cases, have maximum readability. The two do not always go together, but there is a fair body of research into the subject and some combinations have been found more pleasing than others. This deserves study.

Colour harmony   Here again, academic rules of harmony should be eschewed in favour of emotionally satisfying combinations which will add something to the promotional appeal of the package.

3.5   Printing ink

No discussion of colour and packaging would be complete without mention of printing ink. This, in the great majority of cases, provides the colour, whether by way of direct printing on the container or boxes or by means of printed labels. Choice of printing ink may make all the difference between success and failure.

Printing ink has both a visual aspect and a functional aspect. The former provides colour, gloss, print quality and good eye appeal and the latter ensures that the print stands up to wear and that the contents are not spoilt by contamination. The ink must also have adequate light fast qualities and it may be necessary to change colour to secure this.

A disposable package whose primary function is eye appeal does not need light fast inks although it does need inks with good cover strength. On the other hand, a package that is likely to remain on the shelf for long periods must have good quality inks with good light fast qualities. Pigments vary in their light fast qualities and it may be necessary to avoid certain colours because of the weakness of available pigments. Most blues have poor light fast qualities.

It is most advisable to discuss the nature of the packaging with the ink maker. Some inks have poor chemical resistance and cannot be varnished; chrome pigments should not be used for food packaging; greys and fawns are particularly difficult to match. It may be impossible to match colours exactly because of the differing composition of inks.

It is necessary to consider the nature of the stock from which the package is made. An ink of a given colour on one type of material will produce a completely different effect on another type of material. A change in the nature of the stock may make the package look completely different unless the ink is reformulated. Nor is it possible to match a printed carton with, say, a metal box or a printed plastics container without using completely different ink formulations in each case. These points are of particular importance where the package reflects the brand image; a change of stock or a change of ink may make quite a different impression and possibly suggest that old stock is being used. The effect is particularly marked when old and new stock is displayed together. The colour of the stock is also important. Even white, or white coated, board needs care. The exact shade of white is significant; the same ink formulation will look different on, say, a 'blue' white or a 'yellow' white. One plastics manufacturer keeps 390 variations of white on file.

Remember that inks and substrates which are perfectly satisfactory in one kind of light may look quite different in another kind of light. Where colour variations are an important part of a design, samples must always be tested under different types of lighting. An ink formulation designed for, say, daylight may not be suitable for the lighting under which the package will normally be seen.

The usual types of printing ink used include:

- flexible packaging   gravure and flexographic inks
- cartons   litho, letterpress and gravure inks
- outers   water based flexographic inks and moisture set letterpress
- plastics   silkscreen and dry offset inks
- cans   stove enamelling inks
- tubes   litho and letterpress ink

Fluorescent inks are increasingly used in packaging applications but usually have limited light fast qualities. Careful consideration and consultation with the ink maker is recommended.

4   Colour index

4.1   Introduction

There is some difficulty in specifying colour despite the fact that there are numerous colour charts, systems, atlases and collections in common use. The name of a colour is no guide because it varies with the manufacturer or producer, and the appearance of any named colour can vary quite considerably according to the chart or system being used. There is no positive standard to which everyone can refer, and all the systems of colour identification have limitations in practice. The widely used British Standard colours (BS4800:1981) are not really suitable for packaging applications because they do not reflect any fashion element. Although suggestions have been made in this book about colours that might be used, too much reliance should not be placed on colour names.

The index that follows is an attempt to provide a convenient means of identifying hues suitable for packaging applications. It includes colour characteristics grouped together under hue headings, together with a summary of the applications of those characteristics and some notes on recommended uses. The subdivisions of the spectrum used in the index have been devised to solve the problem of identifying hues, and variations of hues, for colour research purposes. Colour names listed are descriptive only and in no sense standards; they are simply intended to help in identifying and explaining the various subdivisions. No special virtue is claimed for this method of classification other than one of convenience, and it derives from experience in the recording of trends. It will be noted that hues have been divided into light, medium and dark variations, partly because these are useful in packaging applications but also because they are essential in the recording of trends. Clear variations have been separated from muted variations in some instances, and in other cases warm variations have been separated from cool variations. Pink has been treated as a separate hue. These subdivisions have been found useful by experience.

The subdivision of each hue into light, medium and dark variations is arbitrary, but a convenient way of making the division is to base it on reflectance values:

- light   reflectance value 50 to 70 per cent
- medium   reflectance value 25 to 50 per cent
- dark   reflectance value 5 to 25 per cent

The grey tints mentioned in the final section are very pale colours with a hint of grey; they might be classified with the muted versions of the basic hues but experience has shown that a separate heading is useful. However, they have little application in the packaging context.

References to likes and dislikes are based on general usage and not on fashion. At any particular time, certain colours may be fashionable although they do not rate highly in polls of preference. Brown was a strong trend colour in the early 1980s but is not generally considered a well liked colour. In any process of colour selection, due weight should be given to current preferences.

4.2   Colour categories

Irrespective of hue, colour can be divided into categories such as hard and soft, and the uses of these categories in the packaging context are shown below. This method of subdivision is also useful in recording trends; there is often a trend to pastel or to bright colours and these influence the variations to be used for packages.

Hard colours   Also called warm colours. Recommended for feature colours for most packages because they have greater attraction value and impact. Where package design features a symbol or logo, the symbol and secondary colours are best in strong, hard, colours.

Soft colours   Also called cool colours. Generally recommended for background in packaging applications except where marketing conditions dictate otherwise. Do not use a soft feature on a hard background.

Light (pale) colours   Light variations may be used for features where circumstances dictate. Pale colours look light in weight. Pastels appeal to young people less than to older people; they also suggest sugar confectionery, and the so-called candy colours are a collection of pastels.

Medium colours   Feature colours should generally be of medium tone and be accentuated by a background having a higher reflectance value.

Dark colours   Not usually recommended for packaging applications except in small areas for contrast purposes or as foils. Darker colours look heavier in weight.

Pure hues   Pure hues of medium tone are recommended for most packaging applications, especially for feature colours.

Modified (sophisticated) hues   Modified hues may be used where something subtle is required but do not generally have maximum impact.

Bright (clear) colours   Make the package look larger and nearer to the eye. Young people like large areas of bright colour such as red, yellow and strong blue. Bright colours are preferred in most packaging applications because they have greater impact on the retina of the eye. As a general rule, use brightness against darkness.

Muted colours   May be used to convey fashion or subtlety but they do not have maximum impact.

4.3   Violet group

- light   lilac, heather, orchid
- medium   mauve
- dark   purple

Colours in this group are not generally recommended for packaging because the group is weak in motivation and does not inspire any favourable impulse.

There is often a short lived fashion trend in favour of violet group colours, and when this exists they might be used for products having a fashion connotation. Violet is essentially a background colour with a fashion note. It can be used where it has an appropriate association with a product or service.

The eye finds difficulty in focusing violet shades and they are best not used for features. It is restful, modifies other colours and mixes well with blue.

Recommended uses

Light   orchid is suitable for the packaging of cosmetics; may be used for confectionery and some bakery products; orchid also for caps and closures of cosmetics.

Medium	may be used for confectionery; recommended for florists
Dark	for high fashion products when in vogue
Avoid	use for products used in the kitchen; use with coffee

Attributes

Age	only appeals to the young in a fashion sense; there is often a strong demand for purple from the young; older people generally prefer paler variations
Associations	high fashion; symbol of excellence; purple has associations with royalty
Fashion	the group generally connotes high fashion but in trend terms generally has a short lived popularity
Impulse	little impact
Markets	no particular virtue except in high fashion markets
Mood	enigmatic and dramatic
Personality	people who like purple are artistic, temperamental, aloof and introspective; those who like lavender are quick witted, vain and refined; those who dislike purple are critical and lacking in self-confidence
Preferences	choice 8 in adults, 6 in children
Products	cosmetics, fashion, florists, some bakery products, grape products
Recognition	little impact
Reflectance	depends on exact shade
Regional	no connotation
Seasons	no connotation
Sex	feminine
Shape	suggests an oval; soft, filmy and never angular
Size	makes images look smaller
Smell	lilac, violet, lavender; purple conjures up exotic odours
Stability	restrained colour; purple suggests influence
Taste	associated with sweet things;
Tradition	associated with the Victorian era and with royalty

Visibility	background colour
Warmth	soft colour, cool, not inviting to the viewer

Functions

Coding	not recommended; difficult to distinguish from blue; only to be used when a large number of colours is necessary
Protection	no special virtue
Readability	white on purple, purple on white, purple on yellow have good readability but are not well liked
Temperature control	no special virtues

Applications

Brand image	special consideration
Export	no special applications known
Food	not generally recommended, but paler shades are linked with sweetness and might be used for confectionery and some baby products
Graphic display	best used as background if used at all
Illustration	depends on circumstances
Safety	no safety applications
Tags and labels	fashion applications only
Television	purple and lilac are difficult to fix

4.4 Blue group

	clear blue subgroup	*muted blue subgroup*
• light	pastel, kingfisher	steel
• medium	mallard, summer	wedgwood, saxe
• dark	midnight, navy	slate

Blue has a perennial appeal to British people. It is an excellent background for food. Blue is a good feature colour for packaging but it has comparatively little shelf impact and is best used for background, especially in graphical applications, because it focuses at a point in front of the retina of the eye.

Blue mixes well with violet and green, but avoid mixing with bluish reds. It is fresh and translucent but may be subduing and depressing if not used with care; smoky blues are particularly subduing. Blue is retiring, dignified and difficult for the eye to focus; while it may have bulk, it does not lend itself to sharpness and detail.

Blue tends to fade quickly and needs care for that reason.

Recommended uses

Light	generally for background; pastel blue for cosmetics and caps and closures for same
Medium	a strong, clear blue is best used for features; sapphire blue has impulse attraction when used for caps and closures; recommended for cosmetics, including aftershave
Dark	midnight blue is a good foil
Avoid	large areas of intense blue
Miscellaneous	a combination of blue/green/silver has a clinical look

Attributes

Age	especially appeals to the young for packages; they like strong blue; otherwise mainly appeals to older people
Associations	the law, coolness, the sky, the navy, the sea, summer, engineering, steel; antiseptic; the symbol of first prize
Fashion	always popular with the British public; muted blues generally indicate high fashion
Impulse	passive and has little impact; best used as background
Markets	good for almost any market, including business men; clear blues are preferred by the mass market
Mood	denotes softness, freshness, cleanliness, the outdoors, but can be subduing if not used with care; muted blues can create a restrained mood and may infer influence
Personality	people who like blue are deliberate, introspective, cautious, conservative; people who dislike it tend to be unstable, neurotic, resentful
Preferences	choice 1 in adults, 6 in children
Products	business, travel, food, fish, steel, engineering, dairy products, electricity, milk; recommended for baby foods and masculine cosmetics such as aftershave
Recognition	poor recognition qualities; difficult to focus
Reflectance	typical light blue 36 per cent, dark blue 8 per cent

Regional	no marked connotations
Seasons	summer, but quite acceptable for all seasons
Sex	male colour, but women buy it because men like it
Shape	circle or sphere; cold, wet, transparent, atmospheric
Size	makes an image look smaller
Smell	associated with antiseptics
Stability	restrained colour, but fresh
Taste	tends to suggest a mild taste or perhaps a watery taste
Tradition	widely used in Georgian times, particularly muted blues; traditionally the symbol of first prize
Visibility	background colour, excellent foil
Warmth	soft colour, cool, not inviting to the viewer

Functions

Coding	a strong, medium blue can be used although it does not have good visibility or recognition qualities; it does provide good contrast with other colours; a clear blue and a muted blue could be used in the same range
Protection	few useful qualities
Readability	blue on white, white on blue, orange on navy blue, navy blue on yellow, yellow on navy blue have good visibility; avoid deep blue on pale yellow, it is disturbing
Temperature control	few useful qualities

Applications

Brand image	special considerations
Export	essentially for temperate climates; the blue of the sky overpowers it in warm climates China: blue and white mean money Sweden: blue and white should not be used Switzerland: conveys textiles Arab countries: blue and white should not be used most countries: conveys detergents popular in the UK and Scandinavia but less so in other countries
Food	best as a background or foil colour but conveys

coolness or cleanliness; paler blues can suggest
that a product is watery; suitable for baby
foods, white vegetables, sea foods, frozen
chicken, tomatoes, dairy products, milk

Graphic display   best used as background; blue and white tend
                    to look insipid under supermarket lighting

Illustration   depends on circumstances

Safety   tends to merge into a grey background but can
                    be seen well at night; tends to make objects
                    smaller and less easy to see; used to convey
                    information such as directions

Tags and labels   a poor colour for tags, although strong blues
                    may be used for children's products and for
                    products sold to men

Television   dark blues tend to look black but most vari-
                    ations reproduce acceptably; ultramarine
                    tends to look turquoise; navy blue and green
                    combinations may present difficulty

4.5   Blue green group

- light   aqua
- medium   turquoise
- dark   Ming, petrol

Similar remarks apply to blue green as for blue, but blue green tends to
appeal to a higher grade market and has a higher fashion image. It
appeals to the young and to women, and has more impact than blue.

It is well recommended for packaging, particularly for food. It is a
good foil colour.

Recommended uses

Light   for cosmetics, particularly those with a clinical
                    theme

Medium   for cosmetics, particularly those used in the
                    bathroom or bedroom
                    turquoise appeals to business men
                    recommended for meat products

Dark   no application

Attributes

Age   appeals to both young and old

Associations	cleanliness, freshness; blue green is the complement of, and flatters, the human complexion
Fashion	turquoise is a high fashion colour
Impulse	stronger blue greens have impulse attraction; otherwise background
Markets	recommended for fashion markets
Mood	coolness, freshness, cleanliness; has more strength than blue
Personality	people who like blue green are discriminating, exacting, sensitive; people who dislike it are disappointed, confused, weary
Preferences	no specific preferences
Products	business, travel, food, cosmetics, meat
Recognition	no special qualities
Reflectance	similar to blue
Regional	no connotations
Seasons	as for blue
Sex	appeals to women, especially aqua
Shape	no special significance
Size	makes an image look smaller
Smell	no particular association
Stability	suggests a higher grade image than blue
Taste	no special associations
Tradition	no special associations
Visibility	turquoise has some properties of green; otherwise background
Warmth	soft colour, cool, more inviting than pure blue

Functions

Coding	turquoise has better qualities than pure blue; a clear turquoise could be contrasted with a muted blue
Protection	no special qualities
Readability	as for blue
Temperature control	no special qualities

Applications

Brand image	a higher quality image than pure blue
Export	similar to blue but has more impact
Food	as for blue; especially good in association with meat
Graphic display	no special connotations
Illustration	depends on application; good where fashion is involved

Safety	no special significance
Tags and labels	better than blue, especially for fashion items
Television	most blue greens tend to look too blue on the screen

4.6 Green group

	clear green subgroup	*muted green subgroup*
● light	pastel, porcelain	sage, opaline
● medium	lime, emerald, apple	olive, jade, quartz
● dark	grass, bottle	forest, fir, georgian

The green group can also be further subdivided into green with a yellow bias and green with a blue bias, but the latter is not recommended for packaging applications and has been ignored in this index.

Green is one of the more difficult colours because there are prejudices against it and it has comparatively little shelf impact. However, it can be recommended for many packaging applications and to create associations. Bright and clear variations are generally recommended; a bright apple green is excellent.

Green is a restful background colour – cool, fresh and soft. It mixes well with yellow, blue and brown but tends to modify other colours. Green is fresh and translucent; pastel and muted greens tend to be subduing, and dark greens are depressive. The colour is restful and kind to the eye but does not provide good contrast and tends to retire; it is not sharply focused and does not lend itself to angularity.

Because of its frequency in nature, green is a big colour and can dominate the eye without disturbing it.

Recommended uses

Light	for business men
Medium	lime green is recommended for cosmetics and better class markets; emerald green for caps and closures, it has a neutral effect; emerald, lime and spring green have a high proportion of yellow and good impulse attraction; medium clear green for farm products, but not for products sold to farmers
Dark	excellent foil; recommended for men's toiletries
Avoid	yellow greens have good visibility but they are not well liked and should not be used for food or for toothpaste, because people do not like

to put it in their mouths; do not use green for childrens' wear (prejudice); cotton tipped swabs in pink and white sold well but not in chartreuse because it reminded people of dirty nappies

Attributes

Age	mainly appeals to older people but light variations can appeal to the young; white/green appeals to the young
Associations	the country, open air, spring, summer, freshness, trees, travel, camping; associated with life, growing crops; there are often religious prejudices against green
Fashion	lime green is a high fashion colour
Impulse	luminous shades of green have good impulse attraction but not dark greens
Markets	there are often prejudices against green and care is necessary; lime green is recommended for better class markets; pastels for the business man
Mood	denotes freshness, restfulness, the outdoors, soothing, refreshing, abates excitement, non-aggressive, tranquil; pastels are subduing, dark greens depressive; a greyish green is influential
Personality	people who like green are good citizens, loyal friends, frank, moral; people who dislike green are frustrated and undeveloped
Preferences	choice 3 in adults, 7 in children
Products	business generally, farm products, garden products, confectionery containing peppermint
Recognition	ranks second in recognition value
Reflectance	typical sage green 27 per cent, dark green 9 per cent
Regional	care is necessary in Ireland and Scotland because of religious prejudice
Seasons	spring, summer
Sex	appeals to both men and women; darker greens appeal to men; olive is masculine
Shape	hexagon; does not lend itself to angularity
Size	little effect on size
Smell	pine, balsam, wintergreen, sage, olive, apple,

trees and shrubs generally; any products which have a woodland or floral scent; associated with peppermint

Stability	safety associations suggest stability; greyish greens are influential; restrained but may be a little depressing
Taste	suggests a lime flavour; menthol in the case of cigarettes; also peppermint
Tradition	light and medium greens distinguish the Adam and Regency periods; moss green is Georgian; British racing green is traditional for cars; white/green is associated with the young; green is also associated with Ireland and with the Catholic faith, and there are often traditional prejudices against it
Visibility	lime green and other yellow greens have good visibility but may not be liked
Warmth	soft colour, neutral, neither inviting nor unfriendly

Functions

Coding	strong greens can be recommended for identification purposes; green has good recognition qualities and yellow greens have good visibility
Protection	green has limited protection qualities and will exclude light with a wavelength of up to 350 nanometres; a green filter will protect against rancidity
Readability	white on green, green on white, red on green, green on red have good qualities
Temperature control	few qualities
Brand image	special consideration
Export	has a general cooling effect and is widely liked; needs to be used with care in Moslem countries because it has religious connotations; also needs care in countries with an Irish element France: associated with cosmetics USA: associated with confectionery China: green hats are said to be a joke Czechoslovakia: green means poison Turkey: a green triangle indicates a free sample
Food	suitable for vegetables, forestry products, baby food, peas, but select the shade which conveys

	the right association with the product; green should be avoided in association with bread, meat or iced cakes, it may suggest mould or bad meat; avoid yellow greens, which may suggest sickness; clear and bright shades are best; greens such as olive are best not used
Graphic display	not sharply focused and does not lend itself to angularity; can dominate the eye without disturbing it; clear greens have most impact
Illustration	good background for any illustration suggesting the country, travel, open air activities and so on
Safety	used to mark first-aid points and the like, in association with white
Tags and labels	good for products intended for men but otherwise no special virtues
Television	dark greens tend to look black but most greens reproduce acceptably; some greens tend to go yellow, others go turquoise; olive is to be avoided; combinations with red and with blue tend to be difficult

4.7   Yellow group

	clear yellow subgroup	*muted yellow subgroup*
• light	lemon, primrose	sand, topaz
• medium	sunshine, amber	gold, mustard
• dark	saffron, maize	antique gold

Yellow is well recommended as a feature colour for packaging, but avoid harsh yellows. It is luminous and has high visibility; however it often has a greenish tinge under fluorescent light and this can cause difficulties, especially in association with food. It is strongly recommended for point of sale and display material because of its high visibility.

Yellow blends well with orange and brown, and also with green to create a 'country' effect. It is sharp, angular and crisp in quality but without substance. Use it sparingly for print read at short range. The eye focuses it clearly without aberration and it stands out well in the dark. It makes objects look larger and tends to advance.

Recommended uses

Light	caps and closures; it has impulse attraction; children's wear
Medium	children's wear
Dark	no application
Avoid	greenish yellow for anything put in the mouth; harsh yellows and mustard which is a modified variation; some versions of yellow orange lack impact; large areas of pale yellow, especially lemon

Attributes

Age	young people like large areas of yellow and yellow appeals strongly to the young, especially the very young
Associations	energy, spring, summer, sunlight, newness
Fashion	variations like golden rod and canary yellow are high fashion and so are subtle shades like antique gold, but it is generally a mass market colour
Impulse	excellent attention getter but avoid harsh yellows
Markets	mass market colour; golden rod is recommended for business men where it is desired to create impact
Mood	energising and conducive to vitality, most cheerful colour; incandescent; pale yellows produce a quieter note
Personality	people who like yellow are intellectual, idealistic, aloof; people who dislike yellow are critical, seeking reality
Preferences	choice 8 in adults, 1 in children
Products	travel, food, photography, cosmetics, lemon drinks, children's wear, butter, cheese, baby foods
Recognition	although it has good visibility, yellow is only third in recognition value and red has more impact on the shelf
Reflectance	a typical medium yellow 50 per cent
Regional	no special connotations
Seasons	spring, summer; yellow brown for autumn
Sex	feminine; stronger shades appeal to men
Shape	triangle or pyramid with the point of apex down

Size	largest of all colours
Smell	vanilla, lemon, honeysuckle, saffron
Stability	suggests vitality but muted shades such as gold create dignity and stability
Taste	too much pale yellow may suggest that a product is weak in flavour
Tradition	gold has always typified richness; traditionally associated with children
Visibility	colour of maximum visibility; canary yellow has a particularly high visibility but golden rod has more impact
Warmth	hard colour, warm, inviting to the viewer; always dominates soft colours

Functions

Coding	excellent visibility but only third in recognition qualities; pale yellow may be difficult to distinguish from white
Protection	amber has useful qualities in protecting products from the effect of light; a greenish yellow protects from rancidity
Readability	black on yellow, yellow on black, blue on yellow, scarlet on yellow, yellow on blue, purple on yellow all have good visibility; deep blue on pale yellow is disturbing
Temperature control	reflects heat well and tends to protect the contents of a package

Applications

Brand image	poor recognition qualities
Export	may be overshadowed by sunlight in tropical climates
	Israel: avoid yellow
	Switzerland: conveys cosmetics
	Eastern countries generally: means plenty but can also mean pornography
	Sweden: gold is not recommended for packages
	Pakistan: saffron and black are the colour of hell
	Buddhist countries: saffron is the colour of priests
	conveys food in most countries except Switzerland

Food	highly visible and recommended to attract attention, but use 'butter' yellow and not harsh variations; gold denotes quality and richness; excellent for children's foods, butter, cheese, baby foods, corn, split peas
Graphic display	strongly recommended for point of sale and display material because of its high visibility; sharp, angular and crisp in quality but lacks solidity; use sparingly for print read at short range
Illustration	according to circumstances
Safety	used to mark hazards, contrasted with black
Tags and labels	black on yellow is highly visible but lacks appeal; good for products sold to children and to men
Television	a satisfactory yellow is not easily created because to 'see' yellow the eye must confuse dots of red and green; at the slightest excuse it goes 'off'

4.8   Orange group

- light   orange
- medium tangerine
- dark   burnt orange

Red orange is the best of all colours for packaging. It is dynamic, pleasing and has high impulse attraction, but too much may be overpowering and may make a container look too heavy. Pure orange lacks the impact of red orange but is good for food packaging. A blackish cast makes orange look dirty; a brownish cast is better.

Orange is good for almost any selling application because it has dramatic emphasis, but use with care. It blends well with yellow brown. It can be trying in large areas. Orange is the best of all colours from a visibility and attention getting point of view.

Recommended uses

Light	for food
Medium	for any packaging
Dark	no application

Attributes

Age	appeals to the young, less so to older people
Associations	autumn, winter, warmth
Fashion	compels interest and is often a fashion colour; good for accent
Impulse	red orange is the best impulse colour of all; it is a colour of great vividness and impact, quite impossible to disregard, and will draw the eye in spite of everything
Markets	secures attention in any market but is a little brash for business use; a mass market colour
Mood	exciting, compels interest and conveys warmth; makes a 'loud sound'; incandescent but, at the same time, hard, dry and opaque
Personality	people who like orange are social by nature, good natured, gregarious; people who dislike orange are serious and cold
Preferences	choice 7 in adults, 5 in children
Products	food, including baby products, bread, cereals, meat, baby foods
Recognition	excellent recognition value
Reflectance	according to shade
Regional	has political connotations in Ireland
Seasons	autumn, winter, especially September and early winter
Sex	probably appeals to women more than men
Shape	no special connotations
Size	makes an image larger
Smell	orange, apricot, tangerine
Stability	brash, not suitable where stability is required
Taste	no special connotations except orange
Tradition	associated with the Protestant faith
Visibility	high visibility, especially red orange
Warmth	hard colour, warm

Functions

Coding	excellent visibility and recognition qualities and high impact
Protection	no special connotation
Readability	black on orange, orange on navy blue, orange on black, all have good readability but little emotional appeal
Temperature control	no special connotation

Applications

Brand image	high impact but a little brash in most cases
Export	no special connotations but use with care in markets with a strong Irish element
Food	red orange is well recommended for many foodstuffs; paler versions are less powerful but equally suitable; may be a little overpowering in large areas; use for bakery products, bread, flour, cereal products, meat, baby food
Graphic display	best of all colours from a visibility and attention getting point of view and ideal from a recognition point of view; for dramatic emphasis and putting over a point
Illustration	generally for accent purposes only
Safety	for marking hazards
Tags and labels	black on orange is highly visible and suitable for mass markets
Television	reproduces well

4.9 Brown group

- light — beige, fawn, buff
- medium — copper, tan
- dark — saddle, chocolate, coffee

Light shades of brown have many packaging applications and have reasonable shelf impact; so do medium tones such as tan. Many browns are the natural colour of foodstuffs, such as chocolate, coffee, nuts. It is essential for fashion applications.

Brown is a 'country' colour, soothing and restful. It blends well with orange and yellow and looks well with green. It is easily harmonised.

Recommended uses

Light	particularly associated with warmth, firelight, and so on; good for most business applications
Medium	tan can be recommended where good taste and refinement are required
Dark	no application
Avoid	for cigarettes in most cases; earthy shades

Attributes

Age	mainly for older people except when in vogue

Associations	autumn, warmth, mellowness, firelight, refinement, quality, the earth
Fashion	lighter shades have a fashion look and are suited to better class trade in upper income groups
Impulse	not a good attention getter; fawn does have some impulse value
Markets	light variations are suitable for upper class markets but not generally for mass markets; yellow brown and buff are acceptable to business
Mood	indicates warmth and mellowness but is hard, dry and opaque; yellow browns create an intense mood; tan is soft and warm and also influential
Personality	people who like brown are conscientious, shrewd, obstinate, conservative; people who dislike brown are gregarious, generous, honest
Preferences	no marked preference
Products	business, cola drinks, soap (but use with care), nuts and products whose natural colour is brown, such as coffee, cocoa
Recognition	no particular virtues
Reflectance	typical light 50 per cent, medium 27 per cent
Regional	no special connotation
Seasons	autumn, winter; fawn and beige for late autumn
Sex	appeals to women, especially lighter versions; yellow browns appeal to men
Shape	no special connotations
Size	lighter shades make an image larger, dark versions have a reverse effect
Smell	coffee, cedar, chocolate, chestnut, cinnamon, ginger, nutmeg, almond
Stability	denotes refinement and quality; tan is recommended for good taste and is influential
Taste	suggests a cola flavour; may connote a rich taste, particularly for coffee
Tradition	most browns belong to the Regency period and the Victorian era
Visibility	neutral
Warmth	hard colour, warm

Functions

Coding	neutral; few qualities to recommend it; lighter versions could be used to contrast with darker colours
Protection	amber glass is used to protect from the effects of light
Readability	not recommended where readability is concerned
Temperature control	no special connotation

Applications

Brand image	special consideration
Export	no special associations
Food	well recommended where there is a relationship with the product, such as coffee, but avoid earthy browns which might suggest dirt; suitable for baked beans, nuts, bakery products, coffee, chocolate
Graphic display	few useful applications except as appropriate background; easily harmonised with most colours
Illustration	special consideration
Safety	no special virtues
Tags and labels	not generally suitable
Television	darker browns are difficult to use and tan may look like pale orange

4.10 Red group

	warm red subgroup	cool red subgroup
• light	pink (see next section)	rose pink (see next section)
• medium	coral, flame, terracotta	rose, claret
• dark	scarlet, tango, cherry	magenta, aubergine

Yellow type reds (warm reds) are generally preferred to cool reds because they have better visibility. Red is an excellent feature colour for packaging, probably the best of all colours. It is pure and has high impulse attraction. It particularly appeals to the young in packaging applications. Vermilion has high visibility, strong emotional and visual impact and appeals to both men and women in all parts of the world; no other colour has so many advantages.

Be careful of shade, preferences change, and a red which is less than pure may appear faded. Red tends to disappear when the eye is dark adapted, such as in a dark cellar, and this may provide difficulties. It should not normally be used as background to a message. It mixes well with pink and blue, but do not mix bluish reds with blue.

Red is sharply focused by the eye and lends itself to sharp angles. It conveys a feeling of durability. It focuses at a point behind the retina of the eye and pulls the image forward. It increases autonomic response. Red is the most easily recognised and identified colour and the easiest to see in daylight conditions. A bright red attracts attention to itself and takes dirt without losing appearance.

Recommended uses

Light	see pink
Medium	flame, vermilion for maximum attraction; strongly recommended for food; for caps and closures with impulse attraction
Dark	maroon is a good foil colour; scarlet is recommended for business markets
Avoid	magenta, it often suggests bad meat; also purplish shades of red generally

Attributes

Age	appeals to both young and old but is particularly good for packaging aimed at the young
Associations	excitement, urgency, warmth, winter, passion, fashion, fertility, fire, the Post Office (in the UK), means 'stop' world wide
Fashion	universally liked
Impulse	yellow reds are excellent attention getters, bluish reds less so; flame and vermilion are particularly recommended
Markets	suitable for all markets; scarlet is recommended for business men
Mood	indicates passion, excitement, anger, warmth; creates a 'loud sound' and is hard, dry and opaque
Personality	people who like it are aggressive, vivacious, passionate; people who dislike are fearful, apprehensive, unsettled
Preferences	choice 2 in adults, 4 in children; almost universally popular irrespective of nationality
Products	business, travel, food, meat products,

	machinery, soap; use for masculine toiletries, but with care for cigarettes
Recognition	first in recognition value and easiest to identify
Reflectance	typical dark red 14 per cent
Regional	no special connotations
Seasons	winter, especially Christmas
Sex	appeals to both sexes
Shape	a square or cube; hard, dry and opaque in ffality, subtle and substantial; lends itself to sharp angles
Size	makes an image look larger and nearer to the eye
Smell	geranium, plum
Stability	bright reds are exciting, cool reds are dignified
Taste	may convey a feeling of richness but can also convey a harsh taste; strawberry
Tradition	deep reds belong to the Victorian era
Visibility	easiest colour to identify and creates most attention but tends to disappear under certain conditions of light, particularly under mercury vapour lighting
Warmth	hard colour, warm, inviting to the viewer

Functions

Coding	easiest of all colours to identify and has excellent visibility but tends to disappear in dim light
Protection	no special virtues
Readability	red on white, white on red, red on green, green on red, red on yellow are the most readable combinations; scarlet is recommended for maximum readability
Temperature control	no particular virtues

Applications

| Brand image | special consideration required |
| Export | universally popular world wide; conveys caution in most countries |

Zambia: use with care, particularly for illustration

China: a happy, prosperous colour

Taiwan: as for China

Czechoslovakia: a red triangle means poison

Food	one of the best of all colours, although red orange is better; warm shades are preferred and have high attraction value; the natural colour of meat
Graphic display	essential for point of sale display work because of its high visibility and impulse attraction; sharply focused by the eye and lends itself to structural planes and sharp angles; best used on a neutral background and should not normally be used as background itself; red letters on a grey ground are more visible than grey letters on a red ground
Illustration	special consideration
Tags and labels	excellent for most tags, especially for products sold to men and children
Television	most variations reproduce acceptably although reds often tend to have an orange cast; maroon is difficult to reproduce and so are combinations of red with green; small variations do not show up satisfactorily; vermilion with an orange cast will simply show up as red

4.11 Pink group

	orange pink subgroup	mauve pink subgroup
• light	pastel pink, peach	rose pink
• medium	coral pink	orchid
• dark	flame pink	—

Pink is similar to red, although it lacks the impact of pure red. It is particularly recommended for cosmetics and fashion promotions. Pink is essentially a feminine colour; it flatters women. It mixes well with red.

Recommended uses

Light	flesh, peach and rose recommended for feminine cosmetics; traditional for baby products; shellfish; peach is particularly good with food; rose appeals to business men
Medium	cigarettes intended for females; caps and closures of cosmetics
Dark	no application

Attributes

Age	appeals to both young and old, but mainly to women
Associations	fashion, flowers, sweetness
Fashion	Subtle pinks are exclusive and essentially feminine
Impulse	luminous tones of pink have good impulse attraction, but not very pale pinks
Markets	for fashion markets or for women
Mood	creates a quieter mood than red and smells 'nice'
Personality	people who like pink are gentle, loving, affectionate; people who dislike pink are resentful and perverse
Preferences	choice 5 in adults, 3 in children
Products	Cosmetics, business, food, confectionery, fish, baby products
Recognition	fair recognition qualities
Reflectance	typical light pink 66 per cent, flesh 51 per cent
Regional	no connotations
Seasons	spring
Sex	appeals to women but not to men; rose pink is particularly feminine but flesh and peach are also good
Shape	no special connotations
Size	neutral
Smell	Carnation, rose, peach, crushed flowers in general
Stability	Creates a gentler mood than red and a feminine atmosphere but does not suggest stability
Taste	associated with sweet things
Tradition	widely used in most periods; traditionally associated with babies, particularly girls
Visibility	no special virtues
Warmth	warm colour, soft

Functions

Coding	not generally recommended except as a contrast to darker colours
Protection	no special virtues
Readability	not recommended
Temperature control	no special virtues

Applications

Brand image	special consideration
Export	no special comments
Food	peach is suitable for many fruits and is a good background for many foodstuffs, particularly sweet things such as confectionery; lacks impact at point of sale
Graphic display	no special virtues
Illustration	special consideration
Safety	no special connotations
Tags and labels	black on pink has good visibility and can be recommended for fashion promotions
Television	a dusty pink against a dark field will look washed out; a brightness ratio of about 5:1 is recommended

4.12   White

White is an excellent foil colour, essentially background. It has maximum contrast with black but can be used with any hue.

White has little visual interest and is difficult to remember and to find. While it can be seen at a considerable distance, the eye has difficulty in seizing upon it. It has little interest because of lack of chromaticity. Blue/white combinations look insipid under supermarket conditions. White looks whiter on black than it does on grey. It should be unquestionably white; if this is not possible use some other colour.

Recommended uses

White is recommended for pharmaceuticals. If used for caps and closures it has a neutral effect.

Attributes

Age	no special age group but young people like large areas of white; white/green is associated with the young
Associations	purity, cleanliness, weddings
Fashion	generally for traditional use, such as weddings
Impulse	little impact
Markets	no special connotations
Mood	no special connotations
Personality	no special connotations

Preferences	choice 4 in adults, 2 in children
Products	pharmaceuticals, sanitary products; right for most products but lacks impact
Recognition	fourth in recognition value
Reflectance	approximately 84 per cent
Regional	no special connotations
Seasons	no special connotations
Sex	neutral
Shape	neutral
Size	Makes the image larger
Smell	no special connotations
Stability	dignified but a little stark and sterile
Taste	no special connotations
Tradition	traditionally for weddings
Visibility	difficult to remember and to find but good background
Warmth	neutral

Functions

Coding	not generally recommended; it has little visual interest but may be used for contrast
Protection	no special virtues
Readability	black on white, white on black, blue on white, white on blue, white on green, green on white, red on white, white on red, white on purple, purple on white, all have good readability
Temperature control	excellent for reflecting heat

Applications

Brand image	special consideration
Export	China: the colour of mourning and should be used with care in illustration; blue and white together mean money
	Hong Kong: not recommended for packaging
	Arab countries: avoid use of white and blue
	Sweden: avoid use of blue and white
	means purity in most Western countries
Food	always correct with food, although it lacks impact; excellent as background; the universal alternative; conveys purity and has been used for health foods to emphasise the health aspect
Graphic display	has little visual interest and is difficult to remember and to find; it can be seen at a

considerable distance but the eye has difficulty in seizing upon it

Illustration	special consideration
Safety	generally used as contrast
Tags and labels	black on white has good readability but lacks interest
Television	may be emphasised by surrounding it with black or dark colours

4.13 Off white group

- cool off whites   ivory, parchment
- warm off whites   magnolia, shell white
- cream   Regency cream

Off white is not generally recommended for packaging except in special circumstances, but can be used instead of white to secure subtle variety.

Attributes

Age	no special age group
Associations	no special connotations
Fashion	some variations are high fashion
Impulse	little impact
Markets	may be used for business markets to create a subtle difference from white
Mood	creates a mood of dignity and safety; gives a distinctive tone
Personality	no special connotations
Preferences	no special connotations
Products	business
Recognition	no special connotations
Reflectance	ivory 66 per cent, cream 68 per cent
Regional	no special connotations
Seasons	no special connotations
Sex	warmer off whites, such as magnolia, appeal to women
Shape	neutral
Size	makes an image larger
Smell	lily of the valley
Stability	creates a more dignified atmosphere than white and gives a distinctive tone
Taste	no special connotations
Tradition	cream belongs to the Victorian era
Visibility	as for white

Warmth   neutral

Functions

Coding   not recommended
Protection   no special virtues
Readability   not recommended
Temperature control   no special virtues

Applications

Brand image   may be used to give a slightly higher image
                          than white
Export   as for white
Food   best avoided in association with food, but
                          warmer off whites might be used as back-
                          ground to suggest softness and quality
Graphic display   as for white
Illustration   Special consideration
Safety   no special connotations
Tags and labels   only for fashion applications
Television   no special connotations

4.14   Grey group

- light   pearl, dawn, oyster
- medium   silver
- dark   charcoal

Grey is not recommended for packaging. It may be used for promotional applications where it is desired to put over a refined and dignified image. It blends well with almost any colour.

Grey is sometimes used to create a historical effect. Charcoal grey is
sophisticated; it is refined and pleasing to the eye. It should be un-
mistakable.

Attributes

Age   not recommended for either young or old;
                          both find it depressing
Associations   dignity, common sense, good taste, high
                          fashion, conservative
Fashion   high fashion, suited to people of good taste
                          and upper income groups
Impulse   no impact
Markets   should not be used for products for the home;

	charcoal appeals to upper class markets but not to mass markets; may be used for business when dignity is required
Mood	creates a mood of dignity, safety and common sense, but in the wrong context can be depressing; a conservative colour which reduces emotional response and is non-committal
Personality	people who like grey are calm, self-contained, sober, dedicated; people who dislike grey are unemotional
Preferences	no special connotations
Products	business
Recognition	little impact
Reflectance	typical light grey 45 per cent, aluminium grey 41 per cent
Regional	no special connotations
Seasons	no special connotations
Sex	appeals to both men and women, but with women it is essentially a fashion shade
Shape	no special connotations
Size	makes an image smaller
Smell	charcoal
Stability	puts over a refined and dignified message and creates a distinctive tone, particularly darker variations
Taste	no special connotations
Tradition	associated with the Victorian era
Visibility	neutral
Warmth	soft colour, cool

Functions

Coding	little to recommend it
Protection	no function
Readability	not recommended
Temperature control	no function

Applications

Brand image	special considerations
Export	no special connotations
Food	not recommended in association with food
Graphic display	may be used where it is desired to put over a refined and dignified image
Illustration	special consideration

Safety   no application
Tags and labels   not recommended
Television   no special problems

4.15   Black

Black is sometimes used in packaging applications to suggest sophistic-ation, but its use needs care. It also indicates high fashion.

   Black is a good foil and provides maximum contrast with white. It is the least visible of colours and make objects look small. It is particularly difficult to see at night. Black should be definitely black.

Recommended uses

Black may be used for pharmaceuticals. It is always safe for lingerie because men like it. For caps and closures, it has a neutral effect.

Attributes

Age   no special connotations
Associations   mourning, high fashion
Fashion   often indicates high fashion
Impulse   no impact
Markets   Often useful in fashion markets; black is bought by men for women
Mood   creates a deep tone.
Personality   people who like black are regal, dignified, passive, sophisticated; people who dislike black are naïve, fatalistic
Preferences   no special connotations
Products   pharmaceuticals, particularly ethicals
Recognition   little impact
Reflectance   5 per cent
Regional   no special connotations
Seasons   no special connotations
Sex   bought by men for women
Shape   no special connotations
Size   the smallest of all colours but gives a sense of weight
Smell   no special connotations
Stability   suggests weight and stability but is a little sombre for most applications
Taste   no special connotations
Tradition   traditionally associated with funerals and mourning

Visibility   least visible of colours
Warmth   cool colour

Functions

Coding   provides maximum contrast with other
                               colours, but not otherwise recommended
Protection   excludes light
Readability   black on yellow, black on white, yellow on
                               black, white on black, black on orange, orange
                               on black have good readability
Temperature control   tends to absorb heat

Applications

Brand image   special consideration
Export   not recommended because it absorbs heat
                               Pakistan: saffron and black are the colours of
                               hell
                               Egypt: associated with evil
Food   few uses except to create contrast; black caps
                               may be used to make a package stand out; not
                               recommended for cake mixes, but acceptable
                               for tea
Graphic display   the least visible of colours, but excellent
                               contrast; stands out well against greyness and
                               provides maximum contrast with white
Illustration   special consideration
Safety   contrasted with yellow or orange to mark
                               hazards
Tags and labels   primarily for the print on tags but may be used
                               for products sold to men and in fashion
                               applications
Television   looks like darkness seen in space

4.16   Grey tints group

- brown   mushroom
- yellow   champagne
- green   mistletoe
- blue   dove

As a general rule grey tints are best not used for packaging because they
lack impact, but they might be selected in some cases where a sophisticated effect is required. The characteritistics will be as for the basic
hues.

Further information

There are comparatively few books which deal with the use of colour in packaging applications. There is a fair volume of published information in the form of articles in the trade and technical press, but it is not easy to refer to them and it has not been thought worth while to try to list suitable references.

Much of the information in this book is based on the researches and experience of the author, and it includes a summary of his articles and papers. Reference has also been made (*inter alia*) to files of the following journals:

Adweekly
Biscuit Maker and Plant Baker
Confectionery Manufacture and Marketing
The Director
The Economist
Financial Times
Food Manufacture
Frozen Food News
The Grocer
Hardware Trade Journal
Hosiery Trade Review
Industrial Advertising and Marketing
International Paper Board Industry
Institute of Packaging Journal
The Manager
Marketing
Marketing Week
Media Week
Packaging

Packaging News
Packaging Review
The Print Buyer
Time Magazine
Technology in Ireland

The following short bibliography may also be found useful:

Birren, F. *Color in Advertising*, Advertising Requirements, USA, 1954.
Birren, F. *Color Dimensions*, Crimson Press, Chicago, USA, 1934.
Birren, F. *Color, Form and Space*, Reinhold Publishing Corporation, New York, USA, 1961 (now Van Nostrand Reinhold Company).
Birren, F. *Color and Human Appetite*, Food Technology, USA, 1963.
Birren, F. *Colour in Package Design*, Sales Appeal, London, 1961.
Birren, F. *Color Psychology and Color Therapy*, McGraw-Hill, USA, 1950. Reissued by University Books, New York, USA, 1961.
Birren, F. *Color in Your World*, Collier Books, New York, USA, 1962 (Crowell-Collier Publishing Company).
Birren, F. *Creative Color*, Reinhold Publishing Corporation, New York, USA, 1961 (see above).
Birren, F. *The Motivational Approach to Package Design*, Material prepared for Sun Chemical Co., New York, USA at various dates.
Birren, F. *New Horizons in Color*, Reinhold Publishing Corporation, New York, USA, 1955 (see above).
Birren, F. *The Printers Art of Color*, Crimson Press, Chicago, USA, 1935.
Birren, F. *Selling Color to People*, University Books, New York, USA, 1961.
Container Corporation of America *Color Harmony Manual* (Ostwald System), 1948.
Danger, E. P. *Using Colour to Sell*, Gower, London, 1968. Reprinted 1983.
Danger, E. P. 'Packaging evaluation', a series of articles in *Confectionery Manufacture and Marketing*, 1962.
Other articles by E. P. Danger have been summarised in the report as a whole.
Dichter, E. *Consumer Motivation*, McGraw Hill, New York, USA, 1964.
Food Manufacturers Federation *Food Labelling Guide* (various editions).
Munsell, A. H. *The Munsell Book of Color*, The Munsell Color Co. Inc., Baltimore, USA, 1961.
Taylor, F. A. *Colour Technology*, Oxford University Press, Oxford, 1962.

US Department of Commerce *The ISCC-NBS Method of Designating Colors and Dictionary of Color Names*, National Bureau of Standards, USA, 1955.

Attention is also directed to the directives of the EEC relating to packaging.

Product index

The following is a list of the products mentioned by name in this book, together with a reference to the section in which the reference appears and a brief note on the subject discussed.

It should be appreciated that the products listed below are not the only products to which the guidelines set out may be applied but simply those that have been specifically named.

Product	Part	Section	Remarks
Aftershave	IV	1.3	blue
	V	4.4	medium blue
Baby foods	IV	5.2	general
	IV	6.3	yellow, green, coral, blue
	V	4.3	violet
	V	4.8	orange
	V	4.4	blue
Baby products	IV	8.2	red, yellow, strong blue
	V	4.11	pink
Baked beans	IV	6.3	brown
	V	4.9	brown
Bakery products	III	17.3	benefit from being seen
	IV	6.3	general
	V	4.8	orange
	V	4.9	brown
	V	4.3	violet
Barrier creams	IV	1.3	masculine colours
Bath salts	IV	1.2	general, gift, luxury
Beans	III	17.3	do not pack in foil
Beer	IV	8.1	brown bottles

Product	Part	Section	Remarks
Beetroot	IV	6.3	dark red
Beverages	III	19.2	PET bottles
Biscuits	IV	5.2	general
	IV	6.3	general
Bleach	III	17.2	fear of leaks
Bread	IV	6.3	orange, red, red orange; avoid green
	V	4.8	orange
Butter	IV	6.3	yellow
	V	4.7	yellow
Cake mixes	IV	6.3	general, avoid black
Cereals	V	4.8	orange
Cheese	V	4.7	yellow
Cherries	IV	6.3	bright red
Chemicals	IV	7	marking hazards
Children's products	V	4.4	blue
Children's wear	IV	4	yellow
	V	4.7	yellow
	V	4.6	green
Chocolates, boxed	IV	2.1	general
Chocolate	V	4.9	brown
Cigarettes	IV	8.5	red, harsh; pink, feminine; brown, mature; green, menthol
	V	4.6	green
	V	4.9	avoid brown
	V	4.10	red
	V	4.11	pink
Cleaning products	V	4.6	green
Clothing	III	17.3	do not crease
Cocoa	V	4.9	brown
Coffee	IV	6.3	colour of can is important; avoid mauve
	III	17.3	benefits from being seen
	V	4.3	not violet
	V	4.9	brown
Cola drinks	V	4.9	brown
Condensed milk	III	17.2	dislike of rimless can
Confectionery	III	16	green, USA
	V	4.3	violet
	V	4.6	green
	V	4.11	pink
Convenience foods	IV	5.1	general
Corn	IV	6.3	bright gold

The principles outlined in this book apply to products whether or not they are mentioned in the index above. In most cases, the products listed are used as examples or are listed because they deviate from normal rules.